Fuzzy Machine Learning Algorithms for Remote Sensing Image Classification

When on a joke we laugh once....

Than why we cry number of times on a problem....

Life is Fuzzy....

Use Fuzzy Logic....

....Remain Crisp....

Fuzzy Machine Learning Algorithms for Remote Sensing Image Classification

Anil Kumar
Priyadarshi Upadhyay
A. Senthil Kumar

CRC Press is an imprint of the
Taylor & Francis Group, an **informa** business

First edition published 2020
by CRC Press
6000 Broken Sound Parkway NW, Suite 300, Boca Raton, FL 33487-2742

and by CRC Press
2 Park Square, Milton Park, Abingdon, Oxon, OX14 4RN

First edition published by CRC Press 2020

CRC Press is an imprint of Taylor & Francis Group, LLC

ISBN: 978-0-367-35571-5 (hbk)
ISBN: 978-0-429-34036-9 (ebk)

Typeset in Palatino
by Lumina Datamatics Limited

Remember........

Life Once........

Birth Once........

Death Once........

To

Gurus who guided us

Students who worked with us

&

Readers of this book

Have Right Choice....

Nothing is Impossible....

I am Possible....

Impossible....

.......On top of every engineering........

do human engineering with specialization in spirituality (aadhyaatmikata)

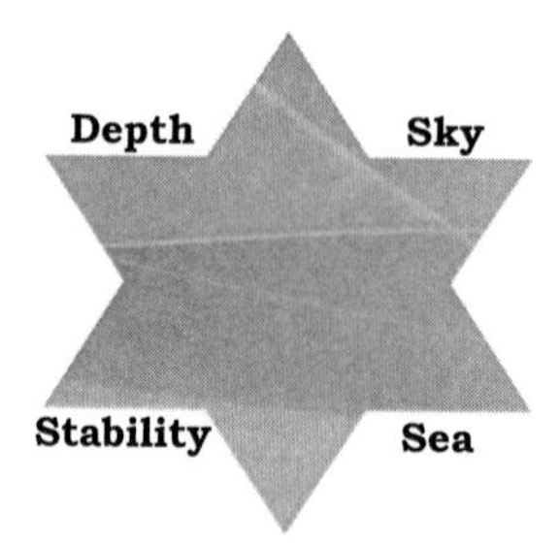

Contents

Foreword

We are living in a rapidly changing world. Buildings serve multiple purposes, agricultural parcels contain various crops, and forests are reduced in size and composition. Spatial information on land use and land cover is coming to us in various ways, and in particular remote sensing images are easily available for that purpose. Their large resources, being really big data, can be freely assessed, but extracting the essential relevant information requires solid and well developed machine learning methods. In addition, we are aware that the world outside is often largely uncertain, or fuzzy. It requires fuzzy machine learning algorithms to respect this uncertainty after classification.

This book contains a comprehensive overview of the latest developments in fuzzy methods for remote sensing image classification. Dr. Anil Kumar together with his colleagues Dr. Priyadarshi Upadhyay and Dr. A. Senthil Kumar have made a remarkable effort to expand and test the available machine learning methods for fuzzy classification. As outstanding experts in their field, they have made the book at a high scientific standard by providing concepts and equations applied to a range of practical studies. In an appendix the book contains the algorithms that are needed in the era of big data for an efficient classification of fuzzy objects.

At this place, I can complement the authors with their achievement and can happily recommend the book to all who are interested in identifying fuzzy information from space, in short: to classify fuzzy objects from remote sensing images.

Dr. IR A. Stein (Alfred)
Professor
Faculty of Geo-Information Science and Earth Observation (ITC)
University of Twente
The Netherlands

This book covers remote sensing image classification from multi-sensor and for different topographic and environmental conditions. Modern machine learning approaches are used for classification into meaningful classes. Traditional approaches, e.g., statistical and fuzzy logic classifiers are covered with wide spectrum. In addition, modern classification approaches based on convolutional neural networks (CNN) with or without long short term memory (LSTM) layers to incorporate recurrent neural network (RNN) flavor are examined with model identification obtained by deep leaning methods. The book is a fresh presentation of the subject; it suits established researchers and graduate students as well.

Dr. Aly A. Farag, Fellow, IEEE and IAPR
Professor of Electrical and Computer Engineering
Director, Computer Vision and Image Processing Laboratory
University of Louisville, Kentucky

Many titles on machine learning are available in the market that cover the field in an excellent manner, like Simon Haykin's *Neural Networks and Learning Machines* and Chris Bishop's *Pattern Recognition and Machine Learning*, among others. However, these titles cover the probabilistic approaches to dealing with the problems of classification and regression, while fuzzy set theory based literature that is increasing in volume is not given its due place in the machine learning space. It is here that this book by Drs. Anil Kumar, Priyadarshi Upadhyay, and A. Senthil Kumar fills an important gap and provides ample complementary material that will immensely benefit the Master's/Doctoral student community and faculty members/researchers.

The material like possibility theory based *c*-means (PCM) and its variants for clustering, hybrid classifiers, temporal data processing and analysis using fuzzy approaches, fuzzy error matrix, etc. are all now available at one place treated in a unified manner by the authors for the benefit of the reader who would have been looking into multiple books/journals for a coverage of these topics. Given the excellent contributions made by the authors to the development of remotely sensed image analysis, the applications of the machine learning techniques in a fuzzy set theoretic framework covered in this title will provide useful insights to the reader that are not easily found elsewhere.

The authors introduce machine learning in the first chapter, followed by ground truth collection in the second chapter. Again, with the exception of Dr. Russell Congalton's book on Accuracy Assessment, the systematic approach to ground truth collection for training the classifier and to test and validate the results covered in this chapter is rarely found elsewhere in detail. Chapter 3 deals with programmed classifiers like FCM, PCM, and others, while Chapter 4 deals with learning based classifiers like artificial neural network (ANN), convolutional neural network (CNN), recurrent neural network (RNN), long short term memory (LSTM), deep learning (DL) principles, etc. Chapter 5 deals with hybrid classifiers where fuzzy set theoretic principles are combined with the learning classifiers.

Accuracy values are routinely reported even in high profile journal publications merely as overall accuracy, kappa coefficient, or user/producer accuracies. There are several other important ways to report accuracy such as fuzzy error matrix and related operators, entropy method, correlation coefficient, receiver operating characteristics (ROC), and a few others. The icing on the cake is the software tool sub-pixel multi-spectral image classifier that has built-in many of the algorithms discussed in the book. This will help the reader explore the algorithms using different datasets and make comparisons with techniques implemented in commercial software packages and in public archives like git-hub.

Overall, this book is a very welcome addition to the machine learning literature, and I recommend that it is seen on the bookshelf of every practitioner/student. Given the expertise of the authors in remote sensing, it is expected that the practitioners of remote sensing related image/data analysis will highly benefit from using this book, but the tools and techniques presented in the book are equally applicable in other domains, and therefore beneficial to students and researchers from those domains.

Dr. Krishna Mohan Buddhiraju
Professor of Centre of Studies in Resources Engineering, IIT Bombay, India

Information on natural resources is vital for the growth and sustenance of human civilization. From time immemorial, this task has been done through very crude methods, such as by walking through an area to data collection through low orbiting Earth Observation (EO) satellite data. Since 1972, satellite data has provided a continuous time series data of the whole world. With the passage of time, the advent of computer and its fast computing characteristics, remote sensing data and its analysis has undergone tremendous change and improvement. The first EO sensors provided data at a very coarse resolution where the pixel size was nearly equal to the size of a football field, whereas in this modern era, the minimum size of the pixel is close to 30 cm.

Earlier, it was thought that finer the resolution of satellite data better would be the information. Every advancement has both its advantages and disadvantages. The interpretation and analysis procedure required an improvement. From simple statistical methods, the trend shifted towards incorporation of the human mind and its thinking process, i.e., artificial neural network to deep learning and now to machine learning and ensemble classifiers. The need for learning has become important. More information means more knowledge and more knowledge means better understanding. Interpretation approaches have changed. Thousands of human hours have been spent to analyze data with different perspective. The ocean of knowledge is getting bigger and it has also become deeper. The remote sensing analysts are looking forward to a scenario where as soon as the data arrives, it is analyzed and presented in the desired form to the user community – an ideal concept. To achieve such a state, it requires high and efficient algorithms to understand the data, correlate to the existing knowledge, and if possible further update the knowledge and improve the learning.

This book, *Fuzzy Machine Learning Algorithms for Remote Sensing Image Classification* being published by CRC press, is one giant leap in learning and understanding analysis, in particular, satellite data. The method of introduction to different approaches reflects the fundamental knowledge of the authors. Each topic has been dealt with utmost care keeping in mind the requirements of the user community. The journey of SMIC has been coolly charted and maneuvered by Dr. Anil Kumar's own career and his constant urge to improve and accommodate new approaches has made it into a power tool. It is time that this software is made available to the user community. Dr. Priyadarshi Upadhyay has been a constant learner in his area of expert and with one of his Guru Dr. Anil Kumar, has embarked on a beautiful journey for quest of knowledge.

I am extremely delighted to see that the two of my rose buds, Dr. Anil Kumar and Dr. Priyadarshi Upadhyay from my garden have now fully bloomed and the fragrance of the same has started to spread worldwide. My best wishes to both of them and may the beacon of knowledge glow brighter in their hands. Dr. A. Senthil Kumar is a well-known person in the area of image processing and my whatsoever interaction with him has always left me more knowledgeable.

Congratulations to all the authors and the book for their efforts and hard work.

Thanks to CRC Press for having boosted the morale of the authors by accepting to publish their work.

GOD bless all of humanity

Dr. Sanjay Kumar Ghosh
Professor and Head
Department of Civil Engineering
Indian Institute of Technology Roorkee
Roorkee, India

'ask not what your country can do for you, ask what you can do for your country'
....John F. Kennedy

Preface

With growing utilization of satellite based earth observation data in environmental and socio-economic management applications, the remote sensing users community faces an uphill task of learning a variety of image classification methods reported in the literature. Unfortunately, description of these classification methods has been found suited to graduate students with computer science or information technology background and not really well suited to a large user community with diverse backgrounds. Moreover, the methods described are largely based on simulated or outdoor images and do not address the complexity and high data handling requirements, typically of satellite images.

The purpose of this book is to cover state of the art image classification techniques in discrimination of earth objects from remote sensing data. In particular, the emphasis is on advances in fuzzy based learning methods for preparing land cover classification maps. The primary emphasis in this book is to provide practical experience to the usage of the technology such as preparation of ground truth data, mono/bi-sensor temporal data processing to study specific crop/vegetation mapping, crop insurance, forest fire mapping, stubble burning, and post disaster damage analysis while incorporating temporal spectral indices databases generated using traditional or class based sensor independent approaches. This book covers from basic methods to the latest ones – fuzzy based machine learning approaches such as convolutional neural networks, recurrent neural networks, etc. – in detail for preparing land cover maps. All fuzzy based machine learning algorithms covered in this book have been included in an in-house developed tool called SMIC: sub-pixel multi-spectral image classifier.

Among the various end products derived from this information, land use/land cover maps have a great impact upon monitoring, planning, and development of resources in individual countries as well as for a region shared by multiple countries. In recent years, the large volumes of remotely sensed data acquired by global space agencies collected through various earth observation sensors with wide application areas are easily available on a global scale. One of the applications of remote sensing data is thematic map preparation. To achieve this, users face challenges like handling mixed pixels, removing isolated pixels or pepper and salt noise, as well as incorporating temporal information for specific class identification. Advancement in machine learning enables an efficient and effective way of extracting geospatial information from remote sensing data. Further, this book will also discuss about some unique applications of specific class mapping such as a specific crop, disaster effected area mapping, etc. with the use of temporal multi-sensor remote sensing data.

Application of fuzzy machine learning algorithms can provide highly realistic classification results while mapping specific classes of interest. These machine learning algorithms can handle mixed pixel problems through a soft computing approach, non-linear modeling of classes while applying the kernel concept and identification of specific classes of interest through the learning algorithm as well as possibilistic based fuzzy classifiers. Independently, the fuzzy classifiers have succeeded in addressing the mixed pixel problem; however, these are not able to resolve the isolated pixel problem, which is then addressed by providing spatial contextual information to the classifier. The neighborhood pixel information in the form of contextual or local information can be incorporated in fuzzy machine learning algorithms to handle noisy pixels as unclassified pixels. Markov random field (MRF) and adaptive local information have been identified as robust techniques to model spatial contextual information.

Chapters in the book are arranged in a manner to explain use of machine learning algorithms for image classification in a systematic manner, to be understood by the professionals and students conveniently. At the end of each chapter, a selected list of references is also given. Chapter 1 provides the introduction on the basics of machine learning algorithms and pattern recognition as well as their capabilities for earth observation. Chapter 2 deals with the importance and necessary ground truth data for training and validation purposes.

Since this book aims at realizing fuzzy based classifiers like fuzzy *c*-means (FCM), possibilistic *c*-means (PCM), noise clustering (NC), as well as hybrid fuzzy classifiers by incorporating spatial contextual information into basic fuzzy based classifier objective function, the middle chapters of the book describe various types of classification algorithms such as fuzzy classifiers, learning based classifiers, and hybrid fuzzy classifiers. Chapter 3 deals with an introduction about clustering algorithms and various types of fuzzy and noise clustering based classifiers. Chapter 4 offers details on the neural and learning based classifiers. It also includes the upcoming machine learning algorithms, like convolutional neural network (CNN), recurrent neural networks (RNN), long short term memory (LSTM), and deep learning which can generate much better classified data from remote sensing images.

Chapter 5 covers various hybrid classifiers generated using the hybridization of methods such as entropy, similarity and dissimilarity, kernel, and contextual with base fuzzy based classifiers. The contextual information has been included through MRF as well as using some local convolution methods.

The classification algorithms discussed in this book are supervised classifiers, therefore exposure about collection of manual training data as well as region growing methods with large number of similarity as well as dissimilarity methods have been covered. The advantage of region growing methods on training data is that the homogeneity in training data increases. The importance of hard as well as soft fraction outputs has also been explained.

Chapter 6 of the book is devoted to use of the multi-temporal data along with the fuzzy based classifiers, as well as deals with the specific land cover extractions. Chapter 7 is concerned with the assessment of accuracy methods for classification. Both relative and absolute methods of accuracy have been discussed. The assessment methods of soft classified outputs through fuzzy error matrix (FERM) with single as well as composite operator have been included as a relative measure of accuracy. The entropy, correlation coefficient, RMSE, and ROC are absolute methods of accuracy.

Appendix A1 is dedicated for the demonstration of SMIC: sub-pixel multi-spectral image classifier, in-house tool, as supervised classifiers. All fuzzy based classifiers mentioned in this book have been incorporated in SMIC. Possibilistic based classifiers have been further implemented in the SMIC tool to process temporal remote sensing data with minimum human intervention to extract specific classes of interest like specific crops, their initial sowing stage mapping or harvesting stage mapping, and application in specific damage area mapping. Appendix A2 covers eleven case studies. These case studies include fuzzy based algorithms studies with similarities measures, non-linearity handling through kernel approaches, controlling noise through MRF and local convolution approaches, and the semi-supervised training approach. Optical and microwave sensors as bi-sensor temporal data approach for paddy fields mapping is also discussed. Specific crop cases included were sugarcane ratoon/plant and paddy burnt fields, and it can be applied to identify wheat, mustard, cassava, ground nut, fennel, cumin crops, etc. as well as post disaster damage identification.

BCD in Geometry of life
Geometry of life has two fixed points,
Birth (B) and Death (D), in between choice (C),
....that's only in our hand....

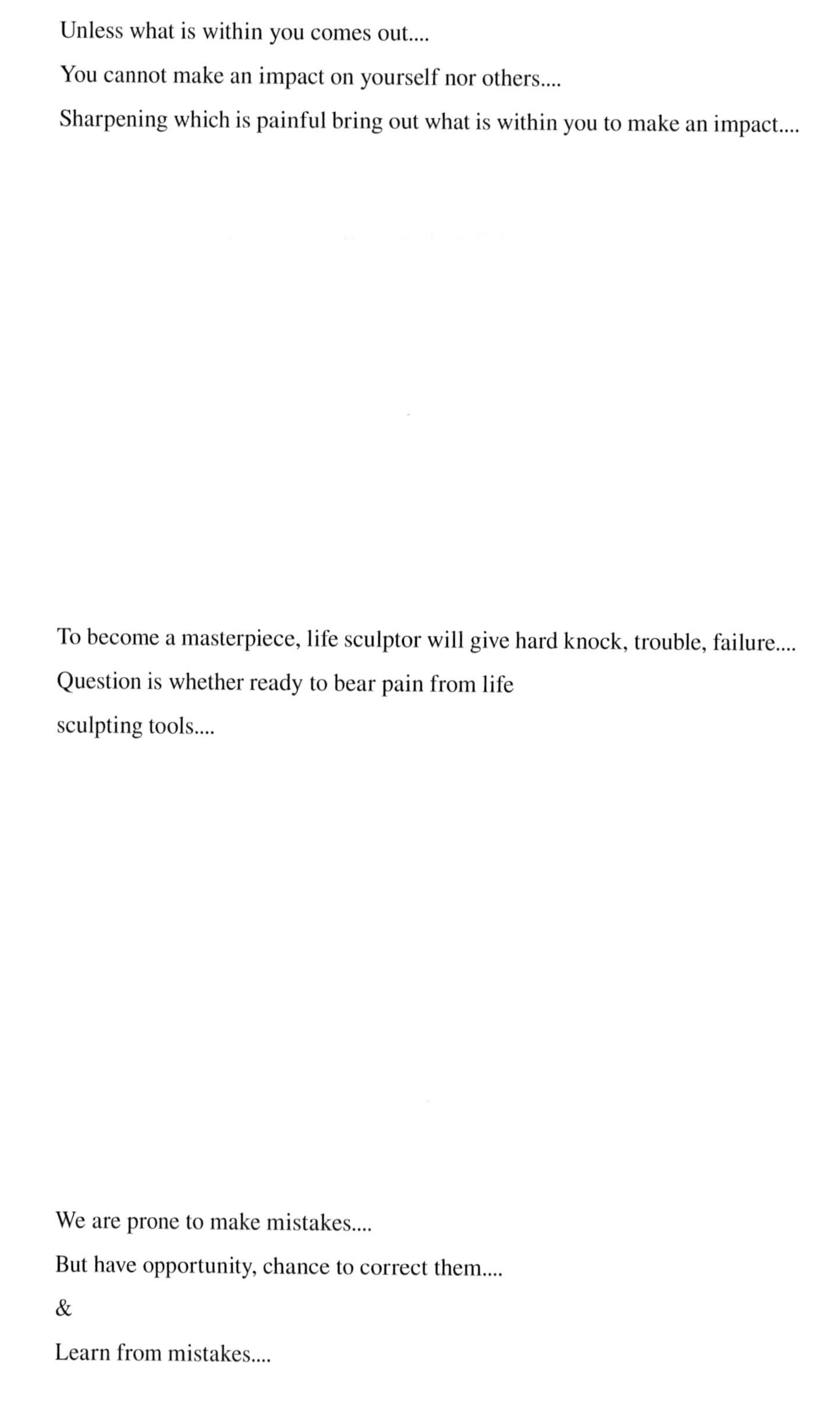

Unless what is within you comes out....

You cannot make an impact on yourself nor others....

Sharpening which is painful bring out what is within you to make an impact....

To become a masterpiece, life sculptor will give hard knock, trouble, failure....

Question is whether ready to bear pain from life

sculpting tools....

We are prone to make mistakes....

But have opportunity, chance to correct them....

&

Learn from mistakes....

Our Gratitude with three R's

Recognizing - Thankful to all those who have made difference in our life!
Remembering to all those who have done good for us - We Mean It!
Reciprocating back for all good done for us.
Thankful to all through three R's of Gratitude
First of all, Praise to God, the creator, destroyer, and the cherisher of the universe.

The thought of writing this book was to share practical experience using fuzzy and learning based classification algorithms in machine learning domain for remote sensing data as well as temporal data from mono-sensor and multi-sensor concepts for specific class identification. In this book, detailed, state of the art mathematical descriptions of fuzzy based as well as learning based algorithms in machine learning domain has been covered.

This book in your hands is the essence of blessings of gurus, elders' support, encouragement, help, advice, and corporation of many people – students, friends, and family members. In the sequel that follows, some names may or may not be there, but their contributions have all been important.

Authors, having been working in ISRO centers, got inspiration encouragement and support from top management starting from the Chairman, ISRO, other dignitaries of ISRO (HQ) and for this we are heartily thankful.

Dr. A. Senthil Kumar sincerely acknowledges Sri. A.S. Kiran Kumar, former chairman, ISRO, for his constant encouragement in pursuing this academic exercise.

We are also thankful to Dr. P. S. Roy, Dr. V. K. Dadhwal, Dr. Y. V. N. Krishna Murthy, and Dr. Prakash Chauhan, leaders of IIRS.

Dr. Priyadarshi Upadhyay is thankful to Dr. M.P.S. Bisht, Director Uttrakhand Space Application Center, and other colleague scientists for constant encouragement of writing this book.

We are thankful to Dr. V. K. Dadhwal for motivated to explore how to incorporate profile information of crop as well as explained how to integrate SAR data with optical data in temporal domain.

We are thankful to Dr. Y. V. N. Krishna Murthy, for learning about clarity and to be crisp in life.

Dr. Anil Kumar and Dr. Priyadarshi Upadhyay are highly grateful to Prof S. K. Ghosh, Head of Department, Civil Engineering, Indian Institute of Technology Roorkee for his constant support and ideas as well as for improvement of writing skills as a supervisor during their PhD. Dr. S. K. Ghosh gave confidence to start PhD research through his statement like; 'smallest step done differently, can be a research'.

Special thanks to Dr. M. Kannan, Sophia Girls College Rajasthan, for motivating through her own experience of writing four books, giving ideas how to write a book, like focusing on small components of chapters and subchapters, and later compiling a full chapter.

The acknowledgment would not be complete without mentioning our family members who provided their love and encouragement. Dr. Anil Kumar thanks his family members, including Mrs. Geeta Kamboj, Ms. Jaya Kamboj, and Mr. Arya Kamboj, Mrs Savita Kamboj (mother) and Dr. J. P. Singh (father), who have suffered from our absence at home during preparation of this textbook.

From Dr. Priyadarshi Upadhyay, special thanks goes to his newborn son, Shreyaan Upadhyay, for taking his precious time while writing the book. He is highly thankful to his lovely daughter Ananya Upadhyay, loving wife Nirja Upadhyay, mother Mrs. Janki Upadhyay, father Mr. Mahesh Chandra Upadhyay and other family members for their every support.

Dr. A. Senthil Kumar specially thanks his wife, Mrs. Senthamarai for her love and care while he was busy in contributing to this book.

Lastly, sincere thanks are due to CRC Press, Taylor & Francis Group, for publishing the manuscript as a textbook as well as the Indian Society of Remote Sensing for considering our book proposal for writing this book. We sincerely acknowledge critical comments received from reviewers, which have brought the content of this maturity. We also thank Dr. IR A. Stein (Alfred), Dr. Aly A. Farag, Dr. Krishna Mohan Buddhiraju and Dr. Sanjay Kumar Ghosh, experts in machine learning area for giving their valuable introductory remarks in the Foreword.

We are thankful to Shri. Gaur Gopal Das, Indian lifestyle coach and motivational speaker who is part of the International Society for Krishna Consciousness, motivated us indirectly to inculcate spirituality and nourishment in this book.

Anil Kumar
Indian Institute of Remote Sensing (ISRO)
Dehradun, Uttrakhand, India

Priyadarshi Upadhyay
Uttrakhand Space Application Center
Dehradun, Uttrakhand, India

A. Senthil Kumar
Centre for Space Science and Technology Education in Asia Pacific (CSSTEAP)
Indian Institute of Remote Sensing
Dehradun, Uttrakhand, India

Have happiness during Journey, not at destination....

Authors

Anil Kumar is a scientist/engineer "SG" and the head of photogrammetry and remote sensing department of Indian Institute of Remote Sensing (IIRS), ISRO, Dehradun, India. He received his BTech degree in civil engineering from IET affiliated to the University of Lucknow, India, and ME degree as well as PhD in soft computing from the Indian Institute of Technology, Roorkee, India. So far he has guided five PhD thesis and six more are in progress. He has also guided several dissertations of MTech, MSc, BTech, and post-graduate diploma courses. He always love to work with PhD scholars, Master and Graduate students for their research work, and motivate them to adopt research oriented professional carrier. He received the Pisharoth Rama Pisharoty award for contributing state of the art fuzzy based algorithms for earth observation data. His current research interests are in the area of soft computing based machine learning for single date and temporal multi-sensor remote sensing data for specific class identification and mapping through in-house development of the SMIC tool. He also works in the area of digital photogrammetry, GPS, and LiDAR.

Priyadarshi Upadhyay is working as a scientist/engineer in Uttarakhand Space Application Centre (USAC), Department of Information & Science Technology, Govt. of Uttarakhand, Dehradun, India. He received his BSc and MSc degrees in physics from Kumaun University Nainital. He completed his MTech degree in remote Sensing from Birla Institute of Technology, Mesra Ranchi, India. He completed his PhD in geomatics engineering from the department of civil engineering, IIT Roorkee, India. He has guided several graduate and masters dissertations in the area of image processing. He has various research papers in SCI listed peer reviewed journals. His research areas are related to applications of time series remote sensing soft computing and machine learning algorithm for specific land cover extraction, microwave remote sensing, and photogrammetry. He is a life member of 'Indian Society of Remote Sensing' and an associate member of 'The Institution of Engineers (India)'.

A. Senthil Kumar earned his PhD from the Indian Institute of Science, Bangalore, in the field of image processing in 1990. He joined ISRO in 1991 and has been serving the Indian satellite programs in various capabilities. His research includes sensor characterization, radiometric data processing, image restoration, data fusion, and soft computing. He is currently the director of UN-affiliated Centre for Space Science and Technology Education in Asia and the Pacific, and former director of Indian Institute of Remote Sensing, Dehradun. He is the president of ISPRS Technical Commission V on Education and Outreach and past Chair and an active member of CEOS Working Group on Capacity Building and Data Democracy. He has published about 110 technical papers in international journals and conferences, besides technical reports. He is a recipient of ISRO Team Awards for Chandrayaan-1, Cartosat-1 missions, outstanding contribution award from Asian Association of Remote Sensing, prestigious Bhaskara Award and Prof. Satiash Dhawan Award conferred by the Indian Society of Remote Sensing.

What is Your Life?

A Candle or Ice Cream....

Candle gives light before it melts....

Ice Cream eaten before it melts....

Dear Esteemed Reader:

Thank you for purchasing this book. We hope this book will enrich you to understand the concepts of fuzzy machine learning for remote sensing image classification. We will be grateful to have your feedback about this book at fuzzymachinelearning@gmail.com

List of Abbreviations

ADFLICM	adaptive fuzzy logic local information *c*-means
ADMPLICM	adaptive modified possibilistic local information *c*-means
ADPLICM	adaptive possibilistic local information *c*-means
AI	artificial intelligence
AIF	adaptive interaction function
ANN	artificial neural network
APF	adaptive potential function
ARVI	atmospherically resistant vegetation index
ASR	automatic speech recognition
AVHRR	advanced very-high-resolution radiometer
AWiFS	advanced wide field sensor
BD	Bhattacharyya distance
CBSI	class based sensor independent index
CMAC	cerebellar model articulation controller
CNN	convolutional neural network
DA	discontinuity adaptive
DL	deep learning
DN	digital number
DNN	deep neural network
DT	decision tree
ED	Euclidean distance
EO	earth observation
EVI	enhanced vegetation index
EVI2	enhanced vegetation index 2
FCM	fuzzy *c*-means
FCME	fuzzy *c*-means with entropy
FAR	false alarm rate
FERM	fuzzy error matrix
FLICM	fuzzy local information *c*-means
GAN	generative adversarial network
GCP	ground control points
GNSS	global navigation satellite system
GPU	graphics processing unit
GUI	graphical user interface
HLN	hybrid learning networks
IPCM	improved possibilistic *c*-means
IRS	Indian remote sensing satellite
JM	Jeffreys-Matusita

KDD	knowledge discovery in databases
KFCM	Kernel based fuzzy *c*-means
KIPCM	Kernel based improved possibilistic *c*-means
KMOD	Kernel with moderate decreasing
KMPCM	Kernel based modified possibilistic *c*-means
KNC	Kernel based noise clustering
KPCM	Kernel based possibilistic *c*-means
LEAST	least operator
LISS-III	linear imaging self-scanning system—III
LISS-IV	linear imaging self-scanning system—IV
LMM	linear mixture model
LP	line process
LSTM	long short term memory
LULC	land use land cover
MAD	mean absolute difference
MAD	median absolute difference
MAP	maximum a posterior
MD	Manhattan distance
MIN	minimum operator
ML	machine learning
MLC	maximum likelihood classifier
MLP	multilayer perceptron
MMD	mean membership difference
MODIS	moderate resolution imaging spectroradiometer
MPCM	modified possibilistic *c*-means
MPCM-s	modified possibilistic *c*-means with constraints
MPCV	mixed pixel class variance
MPLICM	modified possibilistic local information *c*-means
MRF	Markov random field
NBR	normalized burn ratio
NC	noise clustering
NCE	noise clustering with entropy
NDSI	normalized difference snow index
NDVI	normalized difference vegetation index
NDWI	normalized difference water index
NSE	normalized squared Euclidean
OA	overall accuracy
PCM	possibilistic *c*-means
PA	producer's accuracy
PLICM	possibilistic local information *c*-means
PROD	product operator

ReLU	rectified linear unit
RMSE	root mean square error
RNN	recurrent neural networks
ROC	receiver operating characteristics
SA	simulated annealing
SAM	spectral angle mapper
SAR	synthetic aperture radar
SAVI	soil adjusted vegetation index
SCA	spectral correlation angle
SCM	sub-pixel confusion-uncertainty matrix
SID	spectral information divergence
SIM	spectral information measure
SMIC	sub-pixel multi-spectral image classifier
SPOT	Satellite Pour l'Observation de la Terre
SR	simple ratio
SVM	support vector machine
TD	transformed divergence
TP	true positive
TVI	transformed vegetation index
UA	user's accuracy
VI	vegetation index
WRSI	water requirement satisfaction index

When we are beautiful, it's god gift to us….

When we live our life beautiful, it is our gift to god….

Happiness Theorem....

When Divided It Multiples....

1 Machine Learning

1.1 INTRODUCTION

Pattern recognition is a very important field having application in texts, voices, and image-based pattern mapping. In recent times, the automation approaches have been incorporated in place of visual methods. These approaches were initially called clustering or classification. Over time, these pattern recognition approaches have changed in many ways. In the present scenario, these methods are called machine learning approaches and can be supervised or unsupervised. In case of the unsupervised approach, output does not have a label, while the supervised approach has a label in output.

Number of terms such as pattern recognition, machine learning, data mining and knowledge discovery in databases (KDD) etc. show lots of overlap in definitions, and hence are difficult to differentiate. Today machine learning has become a common term for learning methods and originates from artificial intelligence (Nguyen et al., 2019) while KDD is basically for data storage and access, scaling algorithms to massive datasets and interpreting results. Data mining is the practice of examining large, pre-existing databases in order to generate new information (Qiu et al., 2016).

Machine learning develops the ability in machines to learn automatically and further augment through self-experience without much human support or instruction. In machine learning, accessibility to data is provided so that machines learn themselves. Learning of machines requires observations of data such as direct experience or instructions to extract patterns in data to be able to make better decisions. The basic idea is to allow computers to learn on their own, without human intervention, and take action accordingly (Schmidt et al., 2019). Based on training methods, the machine learning algorithm can be categorized as supervised or unsupervised (Musumeci et al., 2019).

1.2 MACHINE LEARNING APPROACHES

Machine learning algorithms are capable to work in supervised mode, generally learned in the past using labeled data to predict future events. In this mode, while working with a known training dataset, the trained learning algorithm handles unknown data to label it. After sufficient training, the system can provide a label to any new unknown input. Machine learning algorithms can compare expected output with actual output and find errors to make modifications in the model accordingly.

On the other hand, unsupervised machine learning algorithms are used when the training samples with labels are not available. An unsupervised learning study tries to divide data into different clusters without labels. The application of unsupervised learning with clustering while processing remote sensing data is generally to know about how many maximum classes can be extracted from given remote sensing images. In the case of learning algorithms, it tries to draw inferences using datasets for describing hidden structures from unlabeled sample data.

The third type of learning algorithm is semi-supervised machine learning algorithms, also called reinforcement learning. As their name suggests, these algorithms come in between supervised and unsupervised learning types. In the semi-supervised approach, a small amount of labeled data and a large amount of unlabeled data are used. These types of learning methods improve system's learning accuracy considerably. Though acquiring large training sample data is costlier and time consuming, the semi-supervised learning approaches are capable to reduce costs and time.

Reinforcement machine learning algorithms produce action and discover errors or rewards while interacting with their environment. The most relevant characteristics of reinforcement learning are trial-and-error search as well as delayed reward. The reinforcement method allows machines to automatically determine the features from the data in order to maximize their performance. To learn which action is a better agent requires a reward feedback, which is known as the reinforcement signal.

Machine learning methods are capable of processing large amounts of data. Further, they provide a faster result with higher accuracy; however, during training it may also require additional time and resources with large amounts of training data for proper training. Integrating machine learning with artificial intelligence and cognitive technologies can bring more effectiveness in processing large volumes of data.

Machine learning, which is implemented as pattern recognition, provides a label to a given unknown input value. This is analogous to statistical based algorithm in statistics. Classification is one of the examples of pattern recognition, which tries to assign each input value to a class while pattern recognition is a broader term which covers a more general problem that includes other types of outputs as well.

So far various categories of machine learning algorithms have been developed such as statistical, fuzzy based, learning based, and statistical cum learning. Statistical based algorithms were developed and applied for thematic map generation extensively in the last few decades. These algorithms have been applied extensively in remote sensing data processing in supervised as well as in unsupervised modes.

Later fuzzy based algorithms came while handling the mixed pixel issues. Mixed pixels are caused by having two or more classes in a particular pixel (Foody 1996a, 1996b). This generally occurs due to mismatch of class size with pixel size or a pixel representing a boundary of classes. Fuzzy based classifiers are of two categories; one follows the probability rule and the other follows possibilistic rules. The former cannot give good results in the presence of untrained classes. In such cases, the possibilistic fuzzy classifiers are successful to handle untrained classes. Other issues come while classes have non-linear separation. This non-linearity between classes can be handled while using the kernel concept in fuzzy based classifiers. Kernels belong to different categories such as local, global, etc. However, this takes data to higher dimensionality while the classes' separation becomes linear. Another issue is off-isolated noisy pixels; these pixels can be handled while incorporating contextual information. It can be achieved through a Markov Random Field (MRF) or with a local convolution approach added in a base classifier such as fuzzy based.

Fuzzy based algorithms use the mean of a class as a reference for criteria calculation which is to be used for classification. While focusing on learning based classifiers, each sample contributes in weight adjustment during training of learning algorithms. Therefore, these learning based algorithms give better results in some cases. Learning based algorithms start like artificial neural network (ANN), convolutional neural network (CNN), recurrent neural network (RNN), and hybrid approaches, etc.

Learning algorithms have various parameters as well as layers to have free parameters. Overall, free parameters can be various layer combinations, layer size, learning rate, momentum, activation functions, types of pooling layers, required epochs, etc (Gu et al., 2018). Therefore, it becomes important to select as well as to optimize the learning algorithm parameters. In the present scenario there are machine learning algorithms belonging to different domains. These can be statistical, tree structure based, fuzzy types, learning based, statistical based, and object based algorithms. Statistical based algorithms are based only on statistical parameters. These statistical parameters can be mean, mode, minimum, maximum, variance–covariance, etc. Very basic statistical algorithms known are *c*-means, parallelepiped, and maximum likelihood classifiers. The *c*-means algorithm works while using the mean vector of a cluster or class. It calculates the distance parameter between unknown vectors to the mean of a cluster/class. The unknown vector has a minimum distance with respect to mean of cluster/class, assigned to that cluster/class. Here cluster means group generated through unsupervised classification approach. Cluster does not have the label of a class. The advantage of unsupervised classification is to find out how many maximum classes can be identified from given images for the same area. But in the case of supervised classification, the label gets assigned to a class which comes from the training or signature data.

A *c*-means algorithm uses only a mean statistical parameter, which is a very simple parameter. This mean parameter can be estimated in an unsupervised approach or can also be calculated in a supervised classification approach. Due to this reason, *c*-means algorithm can be used either as a supervised or as an unsupervised algorithm. The parallelepiped classifier works on the basis of mean as well as standard deviation or minimum as well as maximum parameters of a class. These parameters are used to create a box of a class, and any unknown vector that falls inside the box is labeled to that class. This classifier does not work with a single parameter but requires two different parameters. Since its parameters cannot be estimated through an unsupervised approach, this classifier cannot be used as an unsupervised classifier.

In the case of maximum likelihood classifier (MLC), if distance norm is used as Mahalanobis, then the parameters mean and variance–covariance have to be computed. Due to computation complexity of too many parameters, it cannot be applied as an unsupervised classifier. However, if distance norm is Euclidian in place of Mahalanobis, then MLC can be applied as an unsupervised classifier, as in that case only mean has to be estimated.

In the case of fuzzy based classifiers, most algorithms use only distance measures, so these can also be used as unsupervised classifiers. However, in some cases of the fuzzy based classifiers, final membership values are computed in two steps; therefore, such classifiers can be used as unsupervised classifiers with some lengthy computation.

On the other hand, the decision tree (DT) based classifiers require training data for generating rules; therefore, DT based classifier can be used only as supervised classifiers while learning based classifiers (ANNs, for example) are specific and can be used as either supervised or unsupervised. The specific learning approach in these classifiers works in supervised or unsupervised mode only. Still, most of the learning based classifiers work as supervised mode only.

1.3 UNDERSTANDING PATTERN RECOGNITION

The pattern recognition approach provides a solution for given inputs and performs matching of the inputs while considering their statistical variation. It is unique in comparison to pattern matching approaches, as it finds exact matches in inputs with past existing patterns. Regular expression matching is one of the examples of a pattern matching algorithm, which identifies patterns in textual data. This data is utilized in search capabilities of large text editors and word processors.

Unlike pattern recoginition, pattern matching is not considered as machine learning approach, even after the pattern matching algorithms sometimes provide similar quality output as that provided by pattern recognition approaches. Pattern recognition also can be categorized with respect to the type of learning technique applied to get the output value, whether supervised, unsupervised, or semi-supervised.

There are different terminologies for describing the corresponding supervised and unsupervised learning approaches. Unsupervised classification is also known as clustering, as there is no training data and therefore grouping of input data is into a cluster which is based on some similarity measure. The similarity measures, like distance between instances, can be considered as one of the parameters in place of assigning each input into predefined classes. In some of the application areas, the terminology is different, like in ecology science community, the term "classification" is equivalently used as "clustering." The smallest part of input data for which an output value is generated is formally termed a feature vector.

Some of the pattern recognition algorithms follow probabilistic constraints; these statistical parameters are used for finding the best label for unknown data. Unlike those algorithms, that just give the best label, probabilistic algorithms often provide the probability of the unknown being assigned to the given label. Also, some of the probabilistic algorithms provide N-best labels with their probabilities for N classes, in the place of simply a single best class label. Probabilistic algorithms have the following advantages over non-probabilistic algorithms:

1. Probabilistic algorithms provide a confidence value associated with their choice.
2. Later unknown data can be removed to belong to a class when the confidence of choosing any particular output is too low.
3. Because of the probabilities' nature, partial or complete error propagation can be avoided, more effectively by incorporating probabilities algorithms for larger machine-learning tasks.
4. Feature selection through separability analysis can remove redundant or irrelevant features.
5. Sometimes raw feature vectors transformation techniques can be used as an application of a pattern-matching algorithm.

Pattern recognition is an old term in the field of learning and looks to be relatively obsolete. In the present decade, deep learning in the field of artificial intelligence is a new and popular topic, and machine learning has become a fundamental form of learning, different from pattern recognition and artificial intelligence terms. Machine learning in the present decade is a basic form of learning and the hottest topic is in many start-ups, research labs, and academic areas. Google trends show increase in interest toward deep learning in the present time. A few more points can also be mentioned about the popularity of machine learning terms in the present decade:

1. From the beginning of 2010 onwards, the machine learning buzzword has continuously become popular with various industry and academic applications.
2. In the past few decades, pattern recognition was the hottest topic, but in the present scenario it is declining.
3. In the present decade, machine learning as well as deep learning are being used in various applications. These two buzzwords are new and fast rising areas, overshadowing the popularity of pattern recognition.

1.4 MACHINE LEARNING APPLICATIONS AND EXAMPLES

The machine learning algorithm has been widely applied in day-to-day application areas, such as virtual personal assistants, prediction while commuting, videos surveillance, social media services, email spams, and malware filtering, online customer support, search engine result refining, product recommendations, etc. Further categories of machine learning applications are image and speech recognition, medical diagnosis, prediction, classification, learning association, statistical arbitrage, extraction, and regression. The following examples can better explain the machine learning applications:

1. Spam filtering follows the classification approach, in which the inputs are basically email or other messages and the category of classes are "spam" and "not spam."
2. In regression, it is a supervised problem in which outputs are continuous in place of discrete.
3. In clustering, input data is to be divided into clusters. Further, the labels of groups are not known; that's why it is called an unsupervised approach.
4. In density estimation, distribution of inputs is found in some space.
5. The dimensionality reduction approach brings the data to lower-dimensional space. Topic modeling and principal component analysis are examples.

In place of so many categories of machine learning approaches, learning means to make adjustment in its parameters based on the previous experiences during training process. Developmental learning, elaborated for robot learning, generates its own sequences (also called curriculum) of learning situations to cumulatively acquire repertoires of novel skills through autonomous self-exploration and social interaction with human teachers and using guidance mechanisms, such as active learning, maturation, motor synergies, and imitation.

Pattern recognition methods have been applied in various fields of image acquisition, image enhancement, extraction of classical structure characteristics, binary image processing, and image segmentation of synthetic minerals. It has also been applied for the classification, recognition, testing, and survey of the classical structure characteristics. Once the information is obtained from the image, a complete image description can be created from a two-dimensional image to concrete data and quantitative information on the characteristic structure.

Artificial intelligence (AI) has one of the applications in machine learning (ML), where machines, software, and sensors use cognition. In the present scenario, real working examples of machine learning are:

Virtual Personal Assistants – Siri, Alexa, and Google Assistant are examples of virtual personal assistants. These help in getting information when used through voice.

Predictions while Commuting – Traffic predictions are examples in which congestions are found on some of the routes through ML using online GPS location of vehicles.

Online Transportation Networks – Here booking a cab app estimates the price of the ride, as price surge hours, by predicting the rider demand using ML.

Video Surveillance – Single person monitoring of multiple video cameras is a difficult and boring job. That's why computers are trained through video surveillance systems, powered by AI, to detect crimes while or before they happen.

Social Media Services – Social media platforms use machine learning for their own and user benefits such as: "People You May Know" Here machine learning works on a simple concept considering experiences through user interaction.

- Face Recognition – Users upload a picture with a friend, and this picture is recognized by Facebook using ML.
- Similar Pins – Machine learning is the core element of computer vision, which is a technique to extract useful information from images and videos. Pinterest uses computer vision to identify the objects (or pins) in the images and recommend similar pins accordingly.

Email Spam and Malware Filtering – There are a number of spam filtering approaches to find spam emails. These spam filters are continuously updated and are powered by machine learning.

Online Customer Support – Websites nowadays use Chatbot to reply to customer queries 24 hours a day in place of live representatives; this is possible due to machine learning algorithms.

Search Engine Result Refining – Machine learning is used in Google and other search engines to improve the search results.

Product Recommendations – Once an item is purchased online, the customer continues to receive emails for shopping suggestions, which happens through machine learning.

Online Fraud Detection – Cyberspace has been made secure and monetary frauds can be tracked online through machine learning.

Remote sensing data provides information about classes of great societal benefit. Various areas include urban monitoring, fire detection, or flood prediction from remotely sensed multispectral multi-sensor images, providing great impact on economic and environmental issues. To generate accurate and efficient products from remote sensing data, the multi-disparate data processing has been evolved. Machine learning and signal processing algorithms have played a very important role in remote sensing data processing for extracting meaningful information (Camps-Valls 2009).

Remote sensing has been used for a variety of earth science applications, such as trace gases, retrieval of aerosol products, land surface products, vegetation indices and ocean applications. In remote sensing, machine learning applications are from retrieval of physical variables to bias correction, from code acceleration to detection of diseases in crops and species level classification (Lary et al., 2018).

In recent times, due to the advancement and development of various remote sensing satellites, the application of machine learning is becoming more reliable. Development of high spatial and spectral resolution sensor data is a key factor to increase retrieval and monitoring capabilities. Integration of hyperspectral data analysis with machine learning algorithms has a wide scope in species level identification.

1.5 SUMMARY

Machine learning methods have arisen from pattern recognition. There are various types of the machine learning approaches with a variety of application areas not only in artificial intelligence but also for various earth observation applications. In the next chapter, ground truth data and their application for remote sensing will be discussed.

BIBLIOGRAPHY

Camps-Valls, G., 2009. Machine learning in remote sensing data processing. *IEEE International Workshop on Machine Learning for Signal Processing*, Grenoble, France, 1–6.

Foody, G.M., 1996a. Approaches for the production and evaluation of fuzzy land cover classifications from remotely sensed data. *International Journal of Remote Sensing*, 17 (7), 1317–1340.

Foody, G.M., 1996b. Relating the land-cover composition of mixed pixels to artificial neural network classification output. *Photogrammetric Engineering & Remote Sensing*, 62 (5), 491–499.

Gu, J., Wang, Z., Kuen, J., Ma, L., Shadroudy, A., Shuai, B., Liu, T., et al., 2018. Recent advances in convolutional neural networks. *Pattern Recognition*, 77, 354–377.

Lary, D.J., Zewdie, G.K., Liu, X., Wu, D., Levetin, E., Allee, R.J., Malakar, N., et al., 2018. Machine learning applications for earth observation. In: Mathieu, P.P. and Aubrecht, C. (eds.), *Earth Observation Open Science and Innovation*. ISSI Scientific Report Series, vol. 15. Cham: Springer.

Musumeci, F., Rottondi, C., Nag, A., Macaluso, I., Zibar, D., Ruffini, M. and Tornatore, M., 2019. An overview on application of machine learning techniques in optical networks. *IEEE Communications Surveys & Tutorials*, 21 (2), 1383–1408.

Nguyen, G., Dlugolinsky, S., Bobák, M., Tran, V., López García, A., Heredia, I., Malík, P., et al., 2019. Machine Learning and Deep Learning frameworks and libraries for large-scale data mining: A survey. *Artificial Intelligence Review*, 52, 77–124. https://doi.org/10.1007/s10462-018-09679-z

Qiu, J., Wu, Q., Ding, G., Xu, Y. and Feng, S., 2016. A survey of machine learning for big data processing. *EURASIP Journal on Advances in Signal Processing*, 67, 1–16. doi:10.1186/s13634-016-0355-x

Schmidt, J., Marques, M.R.G., Botti, S. and Marques, M.A.L., 2019. Recent advances and applications of machine learning in solid-state materials science. *npj Computational Materials*, 5, 83. https://doi.org/10.1038/s41524-019-0221-0

When everything is going against, remember airplane takes off against wind....

What stop us to do anything in this life....

Just two words - "What If?"

2 Ground Truth Data for Remote Sensing Image Classification

2.1 INTRODUCTION

In machine learning, reference data samples are a genuine requirement for training as well as assessing the performance of learning algorithms. Based on reference data as seed data, machine learning algorithms can develop cognitive thinking like human beings. Supervised classification is a genuine need of samples required to represent classes for training later testing the classifier. In reinforcement learning also, reference data samples help in developing training data as well as in assessing the performance of the classifiers for dependability of its performance for decision making.

Since remote sensing data covers a large area on the earth's surface, it is impossible to collect true pixels of each class. Also the land cover features vary in space, time, and spectral dimensions. A ground truth pixel is therefore dependent on space (x, y, z), time (t), and spectral (λ) parameters. Unlike the lab environment for machine learning applications, the same earth feature or class, say, vegetation, varies significantly across the image scene. Thus, the representation of true reference data is statistical in nature. The selection of the reference data needs a strong expenditure to ensure an acceptable end result. Ideally, the expert should collect a large number of pixels representing each class randomly across the image frame, which may vary from 10 × 10 km for finer resolution images better than 3 m, to as large as 700 × 700 km for coarser resolution images of 20–100 m pixel size. For applications related to topographic mapping, the reference pixel also includes the altitude or height of the objects.

As the number of images required to be classified is very large for typical requirements like land use/land cover mapping, the ground truth samples are typically extracted from the image itself. This is done by careful extraction of true pixels by field experts who have the knowledge of variability of objects features in optimized locations. Generally, the final set of reference pixels are decided by unsupervised classification followed by visual interpretation with the expert knowledge. So training data is a true representation of each class in the form of pixels, while testing helps to study generalization properties of classifiers or robustness of classifiers for data variability.

In remote sensing, reference data samples refer to truly representing a set of pure pixels of a class. These samples can be used as training as well as testing datasets. The training data is used to generate classification algorithm parameters like statistical as well as tuning of weights for learning based classification algorithms. The term "ground truthing" refers to the process of gathering the proper objective (probable) data for supervised classification and for testing classification results.

In remote sensing, "ground truth" refers to information collected from ground for a class. Ground truth data helps in correlating image information with respect to ground real objects and materials present on that particular location. Collected ground truth information helps in various ways in the field of remote sensing data processing. During the preparation stage of remote sensing data in some deliverable form, the ground truth data helps in calibration, helps in identification, and investigates what is being sensed. Examples include mapping, land use/land cover analysis of remote sensing images, and other techniques in which data are gathered through remote sensing. More explicitly, ground truthing is an activity where pixels from a given satellite image are correlated to what is there on the ground for verifying the content of the pixels on the image. In the context of classified image, ground truth data is used to define the assessment of accuracy.

In remote sensing, domain ground truth is a jargon word, used for near surface observations. In the case of a planetary body, ground truth refers to the on-site gathering of the reference data. It has been utilized to characterize states, conditions, parameters associated with the surface, and any gaseous envelope above it. Main application of collecting ground truth is for calibration and interpretation purposes of remote sensing data as well as training and testing of classification performed. The remote sensing specialist and the beginner should always consider the surface based perspective during all phases of data collection, analysis, and application.

Ground truth, as its name suggests, is usually conducted on the site. It is generally an act of performing ground observations and measuring various properties of the objects. In case of remote sensing, it is done with respect to resolution cells of the digital image. In the present scenario, ground observations are also related to geotagging in the form of geographic coordinates collected using a global navigation satellite system (GNSS) technology. The advantage of geo-tagging of ground truth observation is to have the proper spatial distribution of ground samples. Further, it is also used to overlap these samples on remote sensing datasets. Geo-tagging helps to understand the actual location status of the samples on the ground. Further, geotagging may also help to find out the location errors and how these may affect a particular study.

Ground truth data as training data is used in supervised classification of an image. In the supervised classification approach, the ground truth data is used as training data. It is used to calculate the statistical parameters while using the statistical based classifier. The samples are generally collected through a combination of field work, maps, and field personal experience. The field sample locations are called training sites. In remote sensing classification, the spectral properties of training sites for each class are used to calculate the statistical or weight parameters. Some of the ground truth samples are used as testing data for the assessment of accuracy of the classified thematic maps since different classification algorithms may have varying percentages of errors for same classification data. Thus, it is helpful to identify the best classification algorithms with a given number of classes, while providing the least amount of error.

GNSS observations have been part of ground truth data collection since its beginning by providing the spatial reference to the data. GNSS provides the spatial information tagging to ground truth data while attaching location coordinates. In today's scenario, due to availability of smart phones with GNSS chips and survey grade apps in smart phones, geo-tagging ground truth data can be common to everybody. These apps have facility of accurate positioning, in and around 5–6 m. Specialized smart phone based GNSS receivers location accuracy can reduce the error up to 1–2 m.

Various sources for earth observation data collection are field observations, *in situ* spectral measurements, aerial photographs, detailed project reports and inventory lists, and maps. The requirement of collecting ground truth for supervised classification is for selecting training sites while for unsupervised classification it is for assigning labels to classes. Ground truth can also be conducted to examine the accessibility in different portions of a study area for which classification has to be conducted. When accessibility is limited in the study area as well as limited logistics or circumstances during an off-season, one may think of alternatives like aerial photographs, maps, literature research, interviews with residents, etc. Finally, training data can be prepared while integrating multiple sources of ground data like direct observations, photo documentation, a variety of maps, personal familiarity, etc.

2.2 CREATION OF TRAINING DATA

For a remote sensing image, the number of training sites depends upon the maximum number of classes of interest that can be extracted from a given area. The sites are to be selected based on the fact that they are sufficient in number, homogeneous, truly representing to a class, well-distributed, and class size matching with spatial resolution of the sensor. Thus, the total number of sites depends upon the areal dimensions of the scene to be interpreted.

A good classification is dependent upon the quality of the training data. The quality of training pixels of remote sensing image can significantly affect the performance of the classification algorithm. Campbell (1987) has given criteria for an ideal selection of training data. However, it is not possible to adhere to all the criteria suggested as different classification algorithms require sample training data of different size. Also, the sensitivity of classifiers varies with respect to homogeneity present in training data. The basic purpose of training data is to derive a representative sample for each class (Chen and Stow, 2002). A simple random sampling scheme can be used to select the training pixels from the remote sensing data. At a given confidence level and desired precision, the sample size for training pixels can be calculated using the formula given by Tortora (1978). The sample size depends upon the number of spectral bands used for the classification. In the case of n spectral bands data, the number of training pixels should be greater than $10n$ for each class (Jensen, 1996). Testing samples can be defined while following the Congolton 1991 rule, by saying 75 to 100 samples per class. The sites should be accessible as well as be able to locate on the image. Site location must be near a roadside or open area fields to have good accessibility of classes. In some cases, the classes on the ground are homogeneous and hence training data collected for these classes will be pure, as required. Examples of homogeneous land cover are large water bodies, clouds, snow, desert sand, common types of rocks, specific types of forest, and some core urban areas. Many times, it is difficult to fulfill the criteria of good training samples due to (1) heterogeneity within classes, (2) the "mixed pixel" problem, (3) criteria for selecting number of classes, (4) effect of atmospheric conditions from place to place and on different dates, (5) phenological chances within vegetation/crops, and (6) other unknown variables not able to account.

In remote sensing, another type of ground truthing includes acquiring spectral signatures of different classes using spectrometers, spectrophotometers, and spectroradiometers in the laboratory and/or the field. The samples collected from ground, also called *in situ* measurements, should be brought at the same scale as that of the same radiometry as in remote sensing images.

2.3 CRITERIA FOR GROUND TRUTH DATA

Thus, the role of ground truth data is to compare or correlate remote sensing image pixels with what really is there on the ground. Ground truthing also verifies the numerical value of the pixel on an image with respect to ground object. In addition to classification of images, it allows evolving supervised classification methods to improve accuracy of the classified image and reduce errors in the classification, such as errors of commission and errors of omission.

As mentioned in the previous section, the ground truthing is done in different ways usually on-site or sometimes data collected off-site. The *in situ* may have surface observations as well as measurements of various properties of the objects that are being sensed through remote sensing images. Tagging of ground truth observation with the existing GNSS technology helps to know spatial distribution of ground truth observations. Further these observations can be overlaid on other datasets.

Ground truth data as training data is used to initialize the parameters of classification algorithms. The number of times it can be collected as a combination of field observations, existing maps of study area, and personal experience and are called as training input. In remote sensing, basic classifiers like maximum likelihood classification, parallelepiped classification, and minimum distance classification require statistical parameters. Thus, spectral characteristics from these ground truth observations are used to generate classifier input parameters. These statistical parameters are mean, variance–covariance, standard deviation, minimum, and maximum. Further, ground truth data used as testing samples allow establishing an error matrix which gives assessment of accuracy of the classified output. Different classification algorithms may generate different percentages of error in classified output maps. Therefore, it is important to choose the best classifier among all having least classification error. Moreover, the ground truth data can also be used while performing the atmospheric correction of a remote sensing image.

2.4 TRAINING DATA IN MACHINE LEARNING

The usage of training data in machine learning algorithms is to generate algorithm parameters or adjust weights, which is a part of supervised classification. Thus, training data here act as ancillary data to train (or learn) the classifier. Training data may be called with different names such as training set, training dataset, or learning set. The training sets are samples from which the algorithm parameters are calculated and give an impression, like a computer learns how to process information.

Machine learning algorithms like ANN, CNN, RNN, etc., mimic the ability of the human brain while taking diverse inputs with weights, so to produce activations through individual neurons in the brain. Artificial neurons and their modified versions imitate while providing highly detailed models of how processes in the brain occur. Like in the brain, training data can be defined in different ways for supervised learning algorithms. For sequential decision tree types of algorithms, training data can be in the form of raw text or alphanumerical data. In the case of convolutional neural networks to be applied in the area of image processing and computer vision, the training sets are in the form of large numbers of sub-images called chips. Machine learning algorithms are so complex and so sophisticated that these use iterative training on image chips to generate features, shapes, and even subjects such as people or animals. The training data is required in the supervised approach, which is prior mandatory information.

Machine learning algorithms are capable to learn and predict, once construction and learning of algorithms has been done with given data (Ron and Foster, 1998). These algorithms are capable of data-driven predictions or decisions (Bishop, 2006) once a learning model from input data has been built. Multiple datasets can be used to build the model. Training data (James, 2013) is used to fit the model. This means that model parameters (e.g. weights at connections, learning rate, momentum, hidden layer and its size, combination of convolution max pooling, and LSTM layers) are tuned using a set of examples (Ripley, 1996).

The neural network model is trained with a training dataset through supervised learning methods like gradient descent or stochastic gradient descent. In the process of training, it consists of an input vector and the corresponding output vector, commonly called the target. During the training process, the training dataset works as an input and produces corresponding output, which is then compared with the target value, for each input vector in the training dataset. The model parameters are adjusted while considering the comparison results as well as type of learning algorithm used by the model. Thus, the meaning of model fitting includes both variable selection and parameter estimation.

Once the classification algorithm has been trained (called a fitted model), it can be used to classify the data called a validation dataset (James, 2013). From a validation dataset, an unbiased evaluation of a model fit is done while tuning of parameters has been done using a training dataset (Brownlee, 2017) [e.g. the number of hidden units in a neural network (Ripley, 1996)]. Validation datasets can be used to regularize by early stopping. The early stopping means stop training when the error on the validation dataset increases, as this is a sign of overfitting to the training dataset (Prechelt and Genevieve, 2012). It has been noticed that a validation dataset's error fluctuates during training, generating multiple local minima. Due to this, it has become difficult and necessary to make complicated rules to know when overfitting has truly begun (Prechelt and Genevieve, 2012). Later, the test dataset is used to provide an unbiased evaluation of the trained classified model. If the test data has never been used in training, then the test dataset is also called a holdout dataset.

Thus, in a machine learning algorithm, the training data are the samples used for learning, which means calculating the parameters (e.g., weights, statistical parameters, etc.) of a classifier. Most classification approaches that develop empirical relationships using training data tend to overfit the data, meaning that this algorithm can identify relationships in the training data that do not work for any unknown data in general. This is called the generalization problem.

2.5 VALIDATION DATASET

A validation dataset includes examples used to tune the hyper parameters of a classifier. It is sometimes also called the development set or simply the "dev set," meaning to evaluate the classifier with other type of dataset. In artificial neural networks, hyper parameters are called the number of hidden units. The validation dataset, as well as the testing set, should follow the same probability distribution as the training dataset. The validation dataset, in addition to the testing and training dataset, is required to avoid overfitting during adjustment of classification parameters. The validation is generally required for performance comparison, whereas the performance characteristics, such as accuracy, sensitivity, specificity, F-measure, etc., are done with the testing dataset. The validation dataset thus provides an evaluation of a model to fit for a given training dataset, while adjusting model parameters.

The main objective is always to find out that a particular network should give the best performance with unknown data. This is only possible when comparing various networks, while evaluating the error function with data independent to that used for training. Training of various networks is done while minimizing appropriate error function defined with respect to the training dataset. After training, performance study of networks is conducted through evaluating the error function with an independent validation set. While the network gives the smallest error during validation set are selected and called the hold out method as this procedure may itself lead to over fitting to the validation set. So, through the third independent set of data, called the test set, the performance of the selected network can be confirmed.

2.6 TESTING DATASET

As mentioned in the previous section, a testing dataset is also a ground truth dataset that is different from a training dataset. This means that it should not be used for training of models. Ideally, if a classification model performs fine with the training dataset, it should also work fine with the testing dataset. However, there is much better performance of the classifiers with training dataset, as compared to the testing dataset, which generally indicates overfitting. So actually, a test set is a set of samples applied only to assess the robustness (i.e., generalization) of a fully trained classifier (Ripley, 1996).

Thus, testing data is basically identified to do assessment of accuracy of classified outputs. Some testing data can be used in an assenting way, especially to assess how correctly a given function produces meaningful output from a given input. Other categories of testing data can be applied to assess capability of the function to respond for unusual, extreme, exceptional or unexpected input datasets. Testing data can be generated through a focused or systematic method, as in the case of domain testing or less-focused approaches, having large volume randomized tests. It can also be generated from ground observations, and hence, may be used a number of times. It is not always possible to generate enough testing datasets due to the required time, cost and quality controls.

2.7 SUMMARY

In this chapter, an introduction about the different types of ground truth data and their importance, especially in remote sensing data classification, has been covered. A clear explanation of training validation and testing data has been presented. The following chapter covers various fuzzy based classifiers.

BIBLIOGRAPHY

Bishop, C.M., 2006. *Pattern Recognition and Machine Learning*, p. vii. New York: Springer.

Brownlee, J., 2017. What is the difference between test and validation datasets? Available from October 12, 2017. https://machinelearningmastery.com/difference-test-validation-datasets/.

Campbell, J.B., 1987. *Introduction to Remote Sensing*. New York, London: The Guilford Press.

Chen, D.M. and Stow, D., 2002. The effect of training strategies on supervised classification at different spatial resolutions. *Photogrammetric Engineering & Remote Sensing*, 68 (11), 1155–1161.

James, G., 2013. *An Introduction to Statistical Learning: With Applications in R*, p. 176. New York, London: Springer.

Jensen, J.R., 1996. *Introductory Digital Image Processing: A Remote Sensing Perspective*. 2nd Edition. Upper Saddle River, NJ: Prentice Hall.

Prechelt, L. and Orr, G.B., 2012. Early stopping—But when? In: Mantavaon, G. and Muller, K.-R. (eds.), *Neural Networks: Tricks of the Trade Lecture Notes in Computer Science*, pp. 53–67. Berlin: Springer-Verlag Berlin Heidelberg. https://doi.org/10.1007/978-3-642-35289-8_5

Ripley, B., 1996. *Pattern Recognition and Neural Networks*, p. 354. Cambridge, UK: Cambridge University Press.

Ron, K. and Foster, P., 1998. Glossary of terms. *Machine Learning*, 30, 271–274.

Tortora, R., 1978. A note on sample size estimation for multinomial populations. *The American Statistician*, 32 (3), 100–102.

3 Fuzzy Classifiers

3.1 INTRODUCTION

The thematic map as information is a basic need in resources management and infrastructure development plans. These thematic maps can be prepared in the present scenario using digital image classification techniques. These classification techniques can be of two types, hard or soft approach. Ideally, the hard classification approach is applied when an image has pure pixels. Similarly, soft classification approaches are applied when there are mixed pixels in images. In the real-world scenario, any classifier can be applied as hard or soft. In the following sections, fuzzy based soft classifiers have been presented.

3.1.1 Soft Classifiers

While preparing thematic maps through the digital classification approach, if mixed pixels are present then these pixels cannot be handled through hard classification approaches. Mixed pixel problems can be handled through sub-pixel, or in the present scenario, commonly called soft classification. Before going to discuss soft classifiers, let us understand land cover mapping.

Land cover information is one of the vital components for studying various aspects for global change and environment applications (Sellers et al., 1995). This information of land cover is essential for management of resources, government policies, human activities, etc. (Cihlar, 2000). Thus, land cover is an important determinant of land use and hence, the value of land to the society. This land cover information has important applications in many areas like urban infrastructure planning, post disaster damage assessment and mitigation, agriculture and soil studies, forestry and geoscience studies, and other large application areas.

However, mapping and studying land use/land cover through conventional ground based survey has limitations of resources and time. Firstly, these conventional survey techniques have limitations in monitoring dynamically changing phenomena like crop monitoring or flood inundation. Secondly, for the precise extraction of land cover involves acquiring the spectral data of land surface at various spatial and temporal scales. Multispectral satellite images are used for thematic map preparation using digital classification. In these multispectral images, it is important to have discriminating ground objects. This is possible when these ground objects have unique spectral signatures. Currently, there are very large remote sensing satellites present in earth orbit, such as Landsat, SPOT, IRS, WorldView, etc. Multispectral satellite data based classification is majorly used for the extraction of land cover information. While classifying satellite data for the extraction of objects, every pixel is allocated to a class.

This extraction in general is done by allocating each pixel to a particular representative class; however, in actual scenario, more than one class may exist within a pixel owing to the continuum variation in the landscape as well as the mixed nature of classes (Ju et al., 2003). Such type of pixels in a digital image is referred to as mixed pixels. As a traditional classifier allocates each pixel to one category, hence classification of mixed pixels will not be correct. Further, in a heterogeneous image, such a classifier will accumulate the error.

Mixed pixels in remote sensing images occur due to pixels representing boundaries as well as pixel size not compatible with class size. These mixed pixels have two or more classes. The larger the heterogeneity, the more the chance of mixed pixels. Natural or manmade objects like crop fields, forests, and water are homogenous classes, while urban is a mostly heterogeneous class. The task of the soft classifier is to separate these mixed classes present in a pixel. For resolving the mixed pixel problem, many soft classification techniques have been proposed so far by researchers. The soft classification decomposes a pixel into the class proportions representing membership values. These membership values are stored in fraction output images. A soft classifier can be a statistical classifier like maximum likelihood classifier (MLC), linear mixture model (LMM) (Sanjeevi and Barnsley, 2000; Lu et al., 2004), fuzzy set theory like fuzzy *c*-means (FCM) (Dunn, 1973; Bezdek, 1981; Bezdek et al., 1984), possibilistic *c*-means (PCM), as well as modified versions of possibilistic classifiers (Krishnapuram and Keller, 1993, 1996) and noise clustering (NC) (Dave, 1991), and some are based on support vector machines (SVM) (Vapnik, 1995) and neural networks and their modified versions (Li and Eastman, 2006; Li, 2008).

3.1.2 Traditional Classifiers versus Soft Classifiers

Traditionally, in a hard classification, each pixel is assumed to be composed of only one class. So, the classification technique assigns each pixel to a single class. This is not always true as in reality mixing of land cover occurring within a pixel due to continuum variation in landscape, for example, a low spatial resolution, say 250 m pixel, may contain many land cover classes. Thus, the occurrence of mixed pixels becomes more robust when the classification is either performed at a regional scale or with coarse resolution satellite imagery (Shalan et al., 2003).

In many situations, the conventional hard classification methods like MLC, *c*-means, may tend to either over- or under-estimate the actual area of a land cover and therefore produce the erroneous results. In such cases, a fuzzy concept can be incorporated in classification of mixed pixels which results in multiple and partial class memberships for a given pixel (Pontius and Connors, 2009). Such classification techniques are known as soft classification methods, and the membership value for a particular class indicates the unmixing of information within a mixed pixel. This method allocates a pixel to different classes, in accordance with proportion of their area inside the pixel. The outputs of a soft classification are represented in the form of fraction images equal in number to land cover classes present in the image. According to Ibrahim et al. (2005), a soft classification approach can be helpful for quantifying uncertainties in areas of transition between the different land cover. Therefore, mixed pixels cannot be handled properly by traditional image classifiers. Thus, there is a necessity to develop and implement different soft classification methods.

In this chapter, the current status of some prevalent fuzzy based soft classification methods for land cover has been presented. Among various soft classifiers, fuzzy c-means (FCM), possibilistic c-means (PCM), and noise clustering (NC) are prominent and widely used classifiers. Further, other versions of PCM in the form of improved possibilistic c-means (IPCM) and modified possibilistic c-means (MPCM) have also been introduced. These classifiers have been studied with entropy method, adding contextual methods through MRF as well as neighbor pixel information. These classifiers have also been studied with various similarity/dissimilarity measures. Further kernel approaches have also been incorporated in these classifiers while replacing similarity/dissimilarity measures. All aforementioned classification techniques are based on the basic fuzzy clustering algorithm. According to Foody (2000), these methods have originally been developed as unsupervised classifiers, yet can be modified to supervised mode by providing the class means from the training dataset.

3.1.3 Linear and Nonlinear Classifiers

In a linear classifier, the boundary of separation or separating surface between two classes will be linear or a hyperplane. Examples of such classifiers are logistic regression and the support vector machine (SVM). There can be an infinite number of such hyperplanes. The classification is generally based on the linear combinations of classes. If data are not linearly separable, then classifiers cannot perfectly distinguish the two classes. In such cases, nonlinear functions such as K-nearest neighborhood and kernel based SVM will separate data in higher dimensions to linearize the data (Figure 3.1).

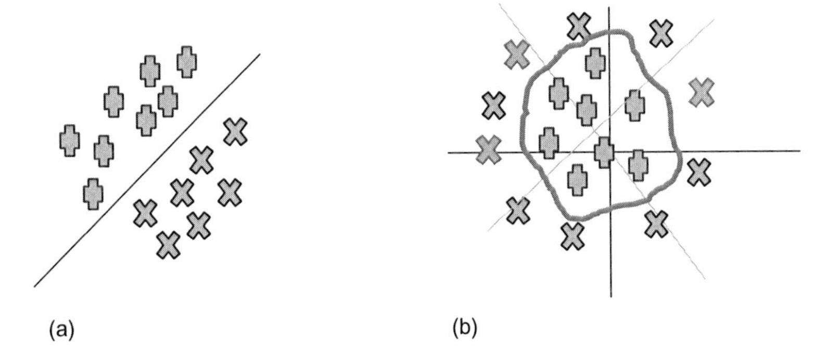

FIGURE 3.1 (a) Linear boundary with classifier hyperplane and (b) nonlinear boundary with classifier hyperplane.

3.2 CLUSTERING ALGORITHMS

In scientific data analysis, there are many situations where it is essential, however difficult, to effectively group similar data, due to scarcity of information about the data. Clustering is a technique for grouping similar data points based on spectral, textural signature details in a cluster rather than their equivalent points in other clusters. It is an unsupervised classification technique (Soman et al., 2006). The potential areas of application of clustering techniques include artificial intelligence, data mining, digital image processing, pattern recognition, and statistics. There is not yet any single clustering algorithm available that can resolve all the clustering problems and perform best for all the datasets. Broadly, there are two types of clustering algorithms: hierarchical and partitional. The basic difference between these two algorithms is that the hierarchical clustering makes a dendrogram structure whereas partitional clustering divides data into a specified number of clusters (Soman et al., 2006).

Another way of classification of clustering algorithms are "hard" and "soft" clustering (Dave, 1991). In the case of hard clustering, each data point (or pixel in image) is allocated to exactly one class, whereas in soft clustering, there will be a fractional allotment for each class for a given data point (Babu and Murty, 1994). Some of the clustering techniques can be generally time-consuming processes; therefore, an ambiguous cluster center or initial seed value is selected for an optimal solution. Many times, clustering algorithms are represented as an optimization problem. In such cases, for hard clustering, the parameter to be optimized is only for the cluster center whereas for soft clustering it includes both the cluster center and membership value for each class. The two basic and popular soft of fuzzy clustering algorithms are fuzzy *c*-means clustering (FCM) and possibilistic *c*-means (PCM). In FCM, the membership value signifies the degree to which a class is shared to a cluster, while in PCM it refers to the belongingness degree (Krishnapuram and Keller, 1993). In comparison to FCM, PCM clustering has been found to be more stable for noise. The outliers or noisy points (or data) are always problematic for effective clustering. These noisy points affect the output of the clustering algorithm, and hence, the outcome is nonrealistic clusters. Further, FCM clustering algorithm makes a certain relation between each data point and a cluster. In other words, each data point is convincingly assigned to a given cluster, whether the point is noisy or not (Dave, 1991). In PCM clustering, according to Krishnapuram and Keller (1993), the problem of noisy points (or outliers) has been resolved somehow by assuming the degree of belongingness of each data point to be equivalent to the membership value. However, the proper handling of noisy points was first proposed by Ohashi (1984) and Dave and Krishnapuram (1997).

3.2.1 Fuzzy *c*-Means (FCM) Classifier

The FCM method was initially introduced by Dunn (1973) and later developed by Bezdek in 1981. It is one of the prevailing clustering algorithm methods for the fuzzy classifications. This method can be applied for partitioning a pixel into different membership values corresponding to the classes present in the digital image.

Each pixel in the image is related to every class by a function known as membership function. The computed values of the membership function are simply known as membership values and vary between zero and one. If the membership value is close to one, then it implies that pixel is strong representative of that class, while membership value close to zero implies that the pixel has weak or no similarity with the information class (Bezdek et al., 1984). Thus, the net impact of such a function is to make fuzzy c-partition of a given data (or satellite image in case of remote sensing). The summation of the all the membership values for each pixel must be equal to unity (Bezdek, 1981). This can be achieved by minimizing the objective function of FCM.

The objective function of the FCM classifier (known as the least square objective function) is given by Equation (3.1) and the distance square is mentioned in Equation (3.2) with constraints in Equation (3.3):

$$J_{\text{fcm}}(U,V)=\sum_{i=1}^{c}\sum_{k=1}^{N}(\mu_{ki})^{m}D(x_k,v_i) \tag{3.1}$$

$$D(x_k,v_i)=d_{ki}^2=\|x_k-v_i\|_A^2=(x_k-v_i)^T A(x_k-v_i) \tag{3.2}$$

where constraints imposed in Equation (3.1) are

$$\sum_{i=1}^{c}\mu_{ki}=1 \quad \text{for all } k \text{ (pixels)} \tag{3.3}$$

$$\sum_{k=1}^{N}\mu_{ki}>0 \quad \text{for all } i \text{ (class)} \tag{3.4}$$

$$0\le\mu_{ki}\le 1 \quad \text{for all } k,i \tag{3.5}$$

where in Equation (3.1)

$U=N\times c$ matrix,

$V=(v_1\ldots v_c)$ is the collection set of vector of information class centers v_i,

μ_{ki} is a class membership value of a pixel,

d_{ki} is the distance in feature space between x_k and v_i,

$D(x_k,v_i)$ is the square of d_{ki},

x_k is a vector (or feature vector) denoting spectral response of a pixel k,

v_i is a vector (or prototype vector) denoting the information class center of class i,

c and N are total number of information classes and pixels respectively,

A is the weight matrix, and

m is the weighting exponent (or fuzzifier) such that $1<m<\infty$. When $m\to 1$, the membership function is hard, and when $m\to\infty$, the memberships are maximal fuzzy (Krishnapuram and Keller, 1993).

The weight matrix A controls the shape of the optimal information class (Bezdek et al., 1984). Generally, it takes the following norm as mentioned in Equations (3.6–3.8):

$$A = I \quad \text{Euclidean Norm} \tag{3.6}$$

$$A = D_i^{-1} \quad \text{Diagonal Norm} \tag{3.7}$$

$$A = C_i^{-1} \quad \text{Mahalonobis Norm} \tag{3.8}$$

where,

I is the identity matrix,

D_i is the diagonal matrix with diagonal elements as eigen values of covariance matrix, and

C_i is given by Equations (3.9) and (3.10):

$$C_i = \sum_{k=1}^{N} (x_k - c_i)(x_k - c_i)^T \tag{3.9}$$

where,

$$C_i = \sum_{k=1}^{N} x_k / N \tag{3.10}$$

After solving the objective function (Equation 3.1), the membership value can be computed as mentioned in Equation (3.11):

$$\overline{\mu}_{ki} = \left[\sum_{j=1}^{c} \left(\frac{D(x_k, \overline{v}_i)}{D(x_k, \overline{v}_j)} \right)^{\frac{1}{m-1}} \right]^{-1}, \text{ where } D(x_k, \overline{v}_j) = \sum_{i=1}^{c} D(x_k, \overline{v}_i) \tag{3.11}$$

where $\overline{\mu}_{ki}$ represents the realization of the class membership value μ_{ki}.

From Equation (3.1), the center of the information class as fuzzy mean $\overline{v}_i$ can be computed as mentioned in Equation (3.12):

$$\overline{v}_i = \frac{\sum_{k=1}^{N} (\overline{\mu}_{ki})^m x_k}{\sum_{k=1}^{N} (\overline{\mu}_{ki})^m} \tag{3.12}$$

where $\overline{v}_i$ represents the realization of the information class center value v_i.

3.2.2 Possibilistic c-Means (PCM) Classifier

The objective function of PCM is derived by adding a new term in the FCM (Krishnapuram and Keller, 1993). The uniqueness of this new term is to emphasize (or assign high membership value) and de-emphasize (or assign low membership value) the representative and non-representative feature points, respectively. In PCM, the untrained classes do not affect the classification outputs (Foody, 2000).

For PCM classifiers, the objective function can be given as in Equation (3.13) (Krishnapuram and Keller, 1993):

$$J_{\text{pcm}}(U,V) = \sum_{i=1}^{c}\sum_{k=1}^{N}(\mu_{ki})^{m} D(x_k, v_i) + \sum_{i=1}^{c}\eta_i \sum_{k=1}^{N}(1-\mu_{ki})^{m} \tag{3.13}$$

It is subject to following constraints Equations (3.14–3.16):

$$\max_{i} \mu_{ki} > 0 \quad \text{for all } k \tag{3.14}$$

$$\text{N} < \sum_{k=1}^{N} \mu_{ki} > 0 \qquad \text{for} \tag{3.15}$$

$$0 \le \mu_{ki} \le 1 \qquad \text{for all } k, i \tag{3.16}$$

where,

η_i is a suitable positive number,

m is a weighting exponent (or fuzzifier) such as $1 < m < \infty$.

According to the first term in Equation (3.13), the distance between the feature vectors and prototype vector should be as low as possible, whereas the second term in the objective function of PCM forces the membership value to be as large as possible. However, interpretation of m is different for FCM and PCM (Krishnapuram and Keller, 1996). Increasing values of m, in the case of FCM, represents increased sharing of pixels in remote sensing images (or data) among all information classes, whereas for PCM, increasing values of m represents the increased possibility of all pixels completely belonging to a given information class. So far, different values of m have been suggested by various researchers for artificial as well as remote sensing data (Krishnapuram and Keller, 1996; Foody, 2000; Ibrahim et al., 2005).

By using the Equation (3.13), the membership values for PCM can be computed as Equation (3.17):

$$\bar{\mu}_{ki} = \frac{1}{1 + \left(D(x_k, \bar{v}_i)/\eta_i\right)^{\frac{1}{(m-1)}}} \tag{3.17}$$

η_i can be computed as Equation (3.18):

$$\eta_i = K \times \sum_{k=1}^{N} \mu_{ki}^{m} D\left(x_k, \bar{v}_i\right) \Big/ \sum_{k=1}^{N} \mu_{ki}^{m} \tag{3.18}$$

where,

$K = 1$, Constant

η_i is also known as the bandwidth parameter (Foody, 2000), and it is a distance at which the membership to a class equals 0.5.

3.2.3 Noise Clustering (NC) Classifier

In the FCM algorithm, the noise data points are clustered with a class having equivalent membership value. Dave (1991) has introduced the NC algorithm to conquer this noise problem. It has been proposed that the outlier or noisy points may be dumped into a separate class or cluster, known as a noise cluster. The quality of clustering analysis does not degrade while applying the NC method. Thus, the main aim of this algorithm is to introduce one additional cluster ($c + 1$) to contain all noisy points.

Figure 3.2 shows an example of a noise clustering algorithm on a given dataset. It results in four valid clusters and few outlier points. These outliers or noisy points belong to a noise cluster. It clearly gives more realistic clusters as compared to PCM. This is owing to the fact that PCM clustering assigns all the data points forcefully (including outliers) to one available cluster, which also affects the cluster center (Dave, 1993).

The objective function for NC can be obtained by adding the term ($c + 1$)th for noise class in Equation (3.1) as Equation (3.19):

$$J_{nc}(U,V) = \sum_{i=1}^{c}\sum_{k=1}^{N}(\mu_{ki})^m D(x_k, v_i) + \sum_{k=1}^{N}(\mu_{k,c+1})^m \delta \tag{3.19}$$

where $U = N \times (c+1)$ matrix and $V = (v_1 \ldots v_c)$.

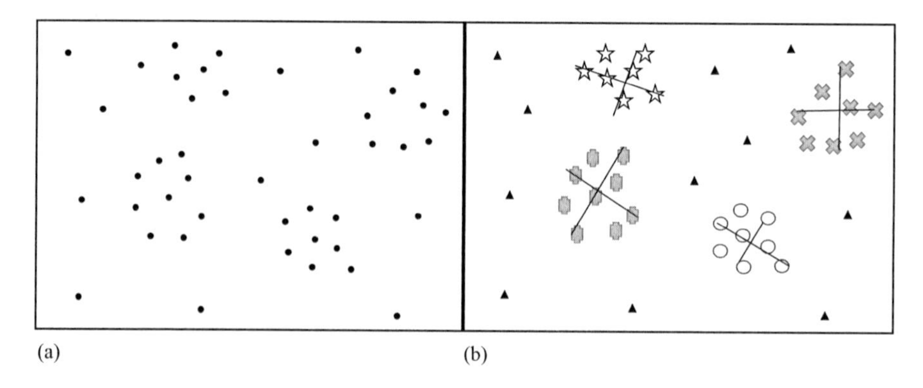

FIGURE 3.2 Noise clustering on a given sample dataset. (a) Noisy data and (b) result of noise clustering.

Noise class has no center and the dissimilarity $D_{k,c+1}$ between x_k can be represented as Equation (3.20) (Miyamoto et al., 2008):

$$D_{k,c+1} = \delta \tag{3.20}$$

where $\delta > 0$ is a fixed parameter.

Equation (3.19) is subject to the following constraints as in Equation (3.21):

$$\mu_f = \begin{Bmatrix} U = \mu_{ki} : \sum_{j=1}^{c+1} \mu_{kj} = 1, \quad 1 \le k \le N; \\ \mu_{ki} \in [0,1], \quad 1 \le k \le N,\ 1 \le i \le c+1 \end{Bmatrix} \tag{3.21}$$

From Equation (3.19), the membership values of information class, noise class, and the cluster mean value can be computed as in Equations (3.22–3.24).

$$\bar{\mu}_{ki} = \left[\sum_{j=1}^{c} \left(\frac{D(x_k, \bar{v}_i)}{D(x_k, \bar{v}_j)} \right)^{\frac{1}{m-1}} + \left(\frac{D(x_k, \bar{v}_i)}{\delta} \right)^{\frac{1}{m-1}} \right]^{-1}, 1 \le i \le c \tag{3.22}$$

$$\bar{\mu}_{k,c+1} = \left[\sum_{j=1}^{c} \left(\frac{\delta}{D(x_k, \bar{v}_i)} \right)^{\frac{1}{m-1}} + 1 \right]^{-1} \tag{3.23}$$

$$\bar{v}_i = \frac{\sum_{k=1}^{N} (\bar{\mu}_{ki})^m x_k}{\sum_{k=1}^{N} (\bar{\mu}_{ki})^m}, 1 \le i \le c \tag{3.24}$$

Noise class is assumed to maintain a static distance from all remaining data points. This static distance is called as noise distance or resolution parameter (δ). Very small values of δ signifies that the majority of data points will be classified as noise data points, while large δ values mean that most of points will be assigned to other classes than noise. Thus $(c + 1)$th class takes the effect of outliers for classification.

Ideally the value of resolution parameter is dependent on the data; however, an approximation of it can be done using Equation (3.25) (Dave, 1991, 1993).

$$\delta^2 = \lambda \left[\frac{\sum_{i=1}^{c} \sum_{k=1}^{N} D(x_k, \bar{v}_i)}{Nc} \right] \tag{3.25}$$

The second term in objective function of NC also ensures that the outliers will get low membership values. Further, constraints in NC are effectively relaxed in a manner that for a given pixel the summation of membership values can also be less than 1, unlike FCM. This opens the door for small membership values to be assigned to noise class and therefore more robust to noise (Dave, 1993, Dave and Krishnapuram, 1997). This constraint is mathematically expressed as Equation (3.26):

$$0 \leq \sum_{i=1}^{c+1} \mu_{ki} \leq 1, \quad 1 \leq k \leq N \tag{3.26}$$

Mathematically, the objective functions of PCM and NC are identical, if the number of cluster (c) is one. However, for $c > 1$, PCM is identical to c separate noise functional to be assigned with a cluster (Dave and Sen, 1997). Further, Equation (3.22) for computing the optimal membership values can be decomposed into two components identical to the equations for computing the membership values of PCM and FCM correspondingly, i.e., $NC_{\mu_{ki}} = FCM_{\mu_{ki}} \times PCM_{\mu_{ki}}$. This equation signifies that NC algorithm is a hybrid of FCM and PCM with their individual qualities inherited (Dave and Sen, 1997).

3.2.3.1 Noise Clustering Algorithm

For an optimal solution in case of the unsupervised mode of NC classification, the following steps should be followed:

1. Select $\delta > 0$, fix m, c, δ, type of A-norm.
2. Compute information class center $\bar{v}_i$ for class $i = 1$ to c using Equation (3.24).
3. Calculate the membership value of information class and noise class using Equations (3.22) and (3.23), respectively.
4. The objective function is calculated using the Equation (3.19).
5. Steps 2 to 4 are repeated until the objective function value converges.

In the case of supervised classification, the noise clustering term can be replaced by the noise classifier due to prior availability of training data.

3.2.3.2 Why Noise Clustering over PCM?

The strength of PCM depends on its mode seeking capability that helps in locating meaningful clusters in the form of dense regions. However, for an optimal performance of PCM, good initialization is required. This can be done using the FCM. Further, relaxation in the membership constraints of PCM results in the independent cluster forming processes. Furthermore, the probability of resulting local minima becomes higher of PCM (Barni et al., 1996).

The noise clustering algorithm, in comparison to its counterparts, is more robust about noise or outliers in the data. It achieves the desired results by making a separate noise cluster for noisy data points. This classifier and its subsequent variations have been proved to have better performance in a noisy environment. Thus, with the ability to incorporate the fuzzy nature of objects as well as allocating a separate noise class, the NC algorithm helps in producing the near realistic classification and results in a noisy environment (Richards, 2013). Moreover, even in the case of untrained classes (i.e., classes which are actually present in the data, but no training data has been given to the classifier), noise clustering does not forcefully allocate the data points to any available class and treats them as noise class only.

3.2.3.3 Drawbacks of Possibilistic *c*-Means (PCM)

For good clustering, the PCM objective function requires some improvement or modification. It is due to some of the drawbacks in the PCM algorithm and these drawbacks are as follows:

1. *Sensitivity toward good initialization*: A good initialization of algorithm affects the outputs in PCM (Barni et al., 1996; Wu and Zhou, 2006).
2. *Problem of coincident clusters*: Due to the independency of columns and rows of typicality matrix, it has the tendency to produce the coincident clusters (Wu and Zhou, 2006).
3. *Neglects the membership:* The typicality factor in PCM reduces the effect of noise, yet it neglects the membership value of class centroid (Wu and Zhou, 2006).

3.2.4 Improved Possibilistic *c*-Means (IPCM)

To overcome the shortcomings of PCM mentioned in section (3.2.3.3), an improved version was proposed by Zhang and Leung (2004). IPCM algorithm is not sensitive to initialization of clusters. This new algorithm is termed as improved possibilistic *c*-means (IPCM). Its advantages over PCM are discussed in the section (3.2.4.1).

3.2.4.1 Advantages of IPCM over PCM

1. *Noise sensitive*: In the presence of noise or outliers, the IPCM method works better than FCM and PCM.
2. *Problem of coincident clusters*: The coincident cluster problem of PCM mentioned in the previous section is resolved by IPCM.
3. *Proper computation of membership and typicality values*: The computed value of typicality and membership matches the intuitive concept, contrary to PCM.

3.2.4.2 Mathematical Formulation of IPCM

For a given dataset $X = \{X_1, X_2, X_3 \ldots X_N\}$ having $1 < c < N$ fuzzy subsets by minimizing the objective function as in Equation (3.27):

$$J_{ipcm}(U,T,V) = \sum_{i=1}^{c} \sum_{k=1}^{N} \mu_{ki}^{m} t_{ki} D(x_k, v_i) + \sum_{i=1}^{c} \eta_i \sum_{k=1}^{N} \mu_{ki}^{m} (t_{ki} \log t_{ki} - t_{ki} + 1) \quad (3.27)$$

where,

$$D(x_k, v_i) = d_{ki}^2 = \|x_k - v_i\|_A^2 = (x_k - v_i)^T A(x_k - v_i)$$

The constraints for the IPCM are:

$$0 \le \mu_{ki},\ t_{ki} < 1 \quad \text{for all } k, i$$

$$\sum_{i=1}^{c} \mu_{ki} = 1 \quad \text{for all } k$$

t_{ki} is the possibilistic value (typicality/compatibility value).

The equations for membership value μ_{ki} and possibilistic value (or typicality/compatibility values) t_{ki} are as in Equations (3.28) and (3.29):

$$\overline{\mu}_{ki} = \left[\sum_{j=1}^{c} \left(\frac{\eta_i \left(1 - \exp\left(\frac{-d_{ki}^2}{\eta_i} \right) \right)}{\eta_j \left(1 - \exp\left(\frac{-d_{kj}^2}{\eta_j} \right) \right)} \right)^{\frac{2}{m-1}} \right]^{-1} \quad \text{for all } k, i, \tag{3.28}$$

$$\overline{t_{ki}} = \exp\left(\frac{-d_{ki}^2}{\eta_i} \right) \quad \text{for all } k, i, \tag{3.29}$$

The cluster center of the information class $\overline{v}_i$ and the bandwidth parameter (or scale parameter) η_i can be computed as in Equations (3.30) and (3.31):

$$\overline{v}_i = \frac{\sum_{k=1}^{N} \mu_{ki}^m t_{ki} X_k}{\sum_{k=1}^{N} \mu_{ki}^m t_{ki}} \quad \text{for all } i \tag{3.30}$$

$$\eta_i = K \frac{\sum_{k=1}^{N} \mu_{ki}^m d_{ki}^2}{\sum_{k=1}^{N} \mu_{ki}^m} \quad K > 0 \tag{3.31}$$

3.2.4.3 Characteristic Features of IPCM

1. IPCM is able to compute both membership values and typicality (possibilistic) values.
2. It can handle the noisy data as outliers and hence, reduce the noise sensitivity.

3. The coincident cluster can be resolved by IPCM.
4. The objective function of IPCM algorithm is flexible enough to incorporate various other types of mathematical norms.

3.2.5 Modified Possibilistic c-Means (MPCM)

The MPCM method is a modified form of PCM for overcoming the shortfalls of FCM and PCM algorithms (Krishnapuram and Keller, 1996; Wu and Zhou, 2008). FCM classifier assigns membership of a pixel across class sum to one with probabilistic constraint. But the membership does not always correspond to the intuitive concept of degree of belonging. Secondly, FCM is sensitive to noise. To overcome these limitations, the PCM algorithm was introduced, as PCM deals with noisy data better. But the drawback of PCM is that it requires good initialization and causes coincident clusters. So MPCM can take care of drawbacks of FCM and PCM and can fit the clusters which are close to each other.

3.2.5.1 Mathematical Formulation of MPCM

For a given dataset $X = \{X_1, X_2, X_3 \ldots X_N\}$ having $1 < c < N$ fuzzy subsets by minimizing the objective function (Wu and Zhou, 2008) in Equation (3.32):

$$J_{mpcm}(U,V) = \sum_{i=1}^{c} \sum_{k=1}^{N} \mu_{ki} D_{ki}^{2} + \eta_i \sum_{k=1}^{N} (\mu_{ki} \log \mu_{ki} - \mu_{ki}) \tag{3.32}$$

where $0 \le \mu_{ki} \le 1$, $D_{ki} = \|x_k - v_i\|$, c is the number of clusters (or classes), N is the number of data points, μ_{ki} is the typicality value of x_k in class i, and η_i is the scale or distribution parameter which depends on all the data which is computed by the Equation (3.33):

$$\eta_i = \frac{\sum_{k=1}^{N} \mu_{ki,fcm}{}^{m} D^{2}{}_{ki}}{\sum_{k=1}^{N} \mu_{ki,fcm}{}^{m}} \tag{3.33}$$

$\mu_{ki,\,FCM}$ is the terminal membership value of FCM. In this case of MPCM, the typicality values (μ_{ki}) and cluster centers (v_i) are obtained as stated in Equations (3.34) and (3.35), respectively, when $D_{ki} = \|x_k - v_i\|$ for all i and $k > 1$, X contains at least c distinct data points, and min $\boldsymbol{J_{MPCM}}(\boldsymbol{U},\boldsymbol{V})$ is optimized.

$$\mu_{ki} = \exp\left(-\frac{D_{ki}^{2}}{\eta_i}\right), \forall i,k\ ;\ (\text{Typicality values for MPCM}) \tag{3.34}$$

$$v_i = \frac{\sum_{k=1}^{N} \mu_{ki} x_k}{\sum_{k=1}^{N} \mu_{ki}}, \forall i\ ;\ (\text{Cluster centers}) \tag{3.35}$$

3.3 SUMMARY

In this section, what is a thematic map and how this can be generated using remote sensing data has been discussed. The presence of mixed pixels in remote sensing images has been mentioned as well as how fuzzy based classifiers can be applied to handle mixed pixels. Basic FCM fuzzy classifiers to the MPCM modified version of fuzzy classifiers has also been discussed. Each classifier's advantages and disadvantages have been mentioned. The next chapter will be on understanding of learning based classifiers.

BIBLIOGRAPHY

Babu, G.P. and Murty, M.N., 1994. Clustering with evolution strategies. *Pattern Recognition*, 27, 321–329.

Barni, M., Cappellini, V. and Mecocci, A., 1996. Comments on "a possibilistic approach to clustering". *IEEE Transactions on Fuzzy Systems*, 4, 393–396.

Bezdek, J.C., 1981. *Pattern Recognition with Fuzzy Objective Function Algorithms*. New York: Plenum Press.

Bezdek, J.C., Ehrlich, R. and Full, W., 1984. FCM: The fuzzy *c*-Means clustering algorithm. *Computers & Geosciences*, 10, 191–203.

Cihlar, J., 2000. Land cover mapping of large areas from satellites: Status and research priorities. *International Journal of Remote Sensing*, 21 (6), 379–387.

Dave, R.N., 1991. Characterization and detection of noise in clustering. *Pattern Recognition Letters*, 12, 657–664.

Dave, R.N., 1993. Robust fuzzy clustering algorithms. *Second IEEE International Conference on Fuzzy System*, San Francisco, CA, IEEE, 1281–1286.

Dave, R.N. and Krishnapuram, R., 1997. Robust clustering methods: Unified view. *IEEE Transactions on Fuzzy Systems*, 5, 270–293.

Dave, R.N. and Sen, S., 1997. Noise clustering algorithm revisited. *Fuzzy Information Processing Society, 1997. NAFIPS'97, Annual Meeting of the North American, 1997*. IEEE, 199–204.

Dunn, J.C., 1973. A fuzzy relative of the ISODATA, process and its use in detecting compact well-separated clusters. *Cybernetics and Systems*, 3, 32–57.

Foody, G.M., 2000. Estimation of sub-pixel land cover composition in the presence of untrained classes. *Computers & Geosciences*, 26, 469–478.

Ibrahim, M.A., Arora, M.K. and Ghosh, S.K., 2005. Estimating and accommodating uncertainty through the soft classification of remote sensing data. *International Journal of Remote Sensing*, 26 (14), 2995–3007.

Ju, J., Kolaczyk, E.D. and Gopal, S., 2003. Gaussian mixture discriminant analysis and subpixel land cover characterization in remote sensing. *Remote Sensing of Environment*, 84, 550–560.

Krishnapuram, R. and Keller, J.M., 1993. A possibilistic approach of clustering. *IEEE Transaction of Fuzzy Systems*, 1, 429–437.

Krishnapuram, R. and Keller, J.M., 1996. The Possibilistic *c*-Means algorithm: Insights and recommendations. *IEEE Transactions on Fuzzy Systems*, 4, 385–393.

Li, Z., 2008. Fuzzy ARTMAP based neuro-computational spatial uncertainty measures. *Photogrammetric Engineering & Remote Sensing*, 74 (12), 1573–1584.

Li, Z. and Eastman, J.R., 2006. The nature of and classification of unlabelled neurons in the use of Kohonen's self-organizing map for supervised classification. *Transactions in GIS*, 10 (4), 599–613.

Lu, D., Mausel, P., Batistella, M. and Moran, E., 2004. Comparison of land cover classification methods in the Brazilian Amazon basin. *Photogrammetric Engineering & Remote Sensing*, 70, 723–732.

Miyamoto, S., Ichihashi, H. and Honda, K., 2008. *Algorithms for Fuzzy Clustering, Studies in Fuzziness and Soft Computing*, vol. 229, pp. 65–66. Springer-verlag Berlin Heidelberg.

Ohashi, Y., 1984. *Fuzzy Clustering and Robust Estimation in 9th Meet.* HollywoodBeach, FL: SAS Users Grp. Int.

Pontius Jr., R.G. and Connors, J., 2009. Range of categorical associations for comparison of maps with mixed pixels. *Photogrammetric Engineering & Remote Sensing*, 75 (8), 963–969.

Richards, J.A., 2013. *Remote Sensing Digital Image Analysis: An Introduction.* Springer-verlag Berlin Heidelberg.

Sanjeevi, S. and Barnsley, M.J., 2000. Spectral unmixing of compact airborne spectrographic imager (CASI) data for quantifying sub-pixel proportions of parameters in a coastal dune system. *Journal of the Indian Society of Remote Sensing*, 28 (2–3), 187–204.

Sellers, P.J., Meeson, B.W., Hall, F.G., Asrar, G., Murphy, R.E., Schiffer, R.A. and Bretherton, F.P., 1995. Remote sensing of the land surface for studies of global change: Models—Algorithms—Experiments. *Remote Sensing of Environment*, 51, 3–26.

Shalan, M.A., Arora, M.K. and Ghosh, S.K., 2003. An evaluation of fuzzy classification from IRS 1C LISS III data. *International Journal of Remote Sensing*, 24 (15), 3179–3186.

Soman, K., Diwakar, S. and Ajay, V., 2006. *Data Mining: Theory and Practice*, New Delhi, Phi Learning Pvt. Ltd.

Vapnik, V., 1995. *The Nature of Statistical Learning Theory.* New York: Springer-Verlag.

Wu, X.-H. and Zhou, J.-J., 2006. An improved possibilistic *c*-Means algorithm based on kernel methods. *Structural, Syntactic, and Statistical Pattern Recognition*, 4109, 783–791. https://doi.org/10.1007/11815921_86.

Wu, X.-H. and Zhou, J.-J., 2008. Modified possibilistic clustering model based on kernel methods. *Journal of Shanghai University*, 12 (2), 136–140. https://doi.org/10.1007/s11741-008-0210-2.

Zhang, J.S. and Leung, Y.W., 2004. Improved possibilistic *c*-Means clustering algorithms. *IEEE Transactions on Fuzzy Systems*, 12 (2), 209–217. https://doi.org/10.1109/TFUZZ.2004.825079.

4 Learning Based Classifiers

4.1 INTRODUCTION

Learning algorithms do not just have a mathematical function like other algorithms. These learning algorithms have a structure with different layers. These layers have different operations on input data to these layers. These layers have small units called neurons or filters, each one passing "high" or "low" signal when its input receives signals, converting input to output data, or generating different features. These neurons of different layers are connected with the neurons of next layers, making a complicated structure. Before final application of these networks, training has to be done in which weights of networks are adjusted and other parameters of networks are optimized. In today's scenario, these networks are of different forms. Further variations in the networks can be generated while using different layer combinations. In this chapter, various types of neural networks and their advantages and disadvantages have been covered with focus on remote sensing data classification.

4.2 VARIANTS OF ARTIFICIAL NEURAL NETWORK (ANN)

There are various types of classification techniques based on algorithms using different parameters. These parameters can be statistical types, learning types, or combination of statistical or learning types. Statistical parameters used by algorithms can be of type such as mean, maximum, minimum, variance–covariance, standard deviations, etc. Learning parameters are weights and combinations of statistical as well as weight parameters. Algorithms using only weight parameters include artificial neural network (ANN), convolutional neural network (CNN), recurrent neural network (RNN), or other network based algorithms as well as hybrid learning networks (HLN). ANN is developed with inspiration from the biological brain but not identical to the brain.

These learning based systems get trained to perform classification by learning from examples, without using class-specific rules for getting tuned. In image classification, learning algorithms are trained from sample images containing class samples as well as labels, and store the trained learning parameters of the classification algorithms. Using these trained parameters, algorithms can be applied to classify new images.

The ANN structure is made up of layers containing neurons called artificial neurons, and these neurons of each layer are connected with the next layer of ANN (Figure 4.1). This structure of ANN tries to connect layer-wise neurons like in the biological brain. Each input can transmit a signal from one artificial neuron to another coming through the synapses. An artificial neuron, while receiving a signal, processes this input, and then this processed output is passed to artificial neurons present in the next layer.

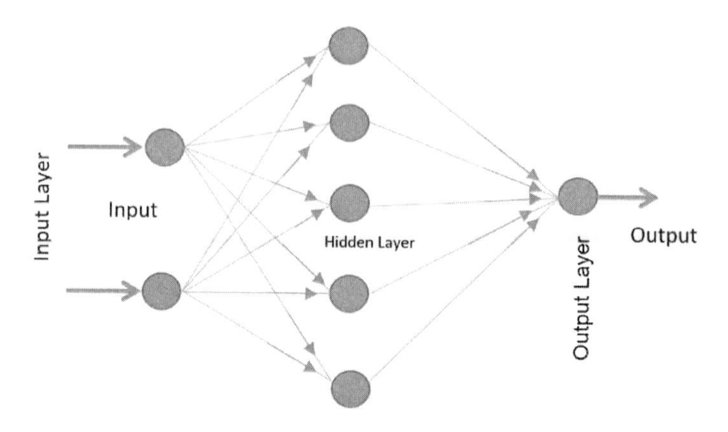

FIGURE 4.1 ANN structure with various layers.

The input coming to each neuron of ANN is a real number. Output from each artificial neuron is computed by activation function with weighted sum of its inputs. Edges (arrows) in ANN provide connectivity between two consecutive layers of artificial neurons. Edges typically have a weight, and these weights are adjusted during the learning process of ANN. The weight adjustment is as per error contributed by that weight. This weight correction can be positive or negative as per adjustment required per weight in each connection. A threshold can be applied on each artificial neuron, so that output from each neuron is only forwarded if the output is more than the threshold defined. Multi-layer perceptron concept is named due to artificial neural networks having multiple layers. The flow of input data goes from the first layer (the input layer) to hidden layers and to the last layer called the output layer.

Artificial neural networks have specific capability, which makes them different in comparison to traditional classifier techniques. The two main benefits to apply neural networks are that they solve problems step-by-step, and they work like expert systems. The behavior of learning algorithms such as ANN, CNN, RNN, etc. is different in comparison to conventional mathematical function based classifiers. To study these intrinsic variabilities present in each class, sufficiently large training or reference data sets are to be fed for training. If the variabilities are large for each class, it is natural that training time is proportionately larger to find an optimal solution. Hence learning process is mathematically a multi-parameter optimization problem.

Artificial neural networks require large training data sets, otherwise their accuracy drops. They have been applied in various applications. Paola and Schowengerdt (1995a, 1995b) delivered an inclusive assessment of the application of the multi-layer perception in remote sensing. In today's scenario, the data can be from disparate sources, due to which data cannot be linear, stationary, or oriented and cannot be modeled comfortably. Neural networks do not require any information about the problem type or statistical distribution of the data set. Initially developed conventional classifiers require that input data should follow a standard distribution pattern, but for neural network, this is not necessary. Learning through which parameters of algorithm is calculated is a first step in a supervised classifier. In the case of learning algorithms, learning scheme such as adjustment of weights has an altogether different meaning. In the case of networks, learning takes place with back-propagation algorithms, etc. After the network has been trained, the next step is to identify the unknown vector (image pixels) belong to which class. Training can be done in different ways, through different learning examples, learning instructions, and learning procedures. A static network is a system which follows clear-cut knowledge and manufacture phases. Networks which learn continuously during processing of data are known as energetic systems.

A learning model can be implemented in three ways the supervised, unsupervised, or a hybrid approach. The learning approach is when reference data is used in the neural network. Supervised and unsupervised training are combined in the hybrid learning methods. To train the system, learning criterion is a type of model that can be used. During iteration of each training, interneuron synaptic weights are updated using specific mathematical processes of learning algorithms. A variety of possible learning algorithms can be used under each learning rule. With a single learning rule, most of the algorithms can be used. While working with supervised or unsupervised learning paradigms using rules and learning algorithms, both will generate different effects. Unlabeled data, i.e., semi-supervised learning data, is also taking advantage in overcoming the small training data problem. This approach has been achieved through pseudo-label (Lee, 2013) and incorporating generative models, such as generative adversarial networks (GANs) (Salimans et al., 2016). But how much these techniques can improve the performance of deep learning (advance form of ANN) is not clear and is an area of active investigation.

Overtraining of learning algorithm occurs when there are too many training examples, as well as when training cycles called epochs are also very large and do not perform well with respect to generalization. Overtraining also occurs when computation in the form of iteration is very large in comparison to dimensionality of the input space. Large numbers of training samples should not be used during training. Different results with respect to quality and robustness of network can be seen with variation in training samples.

While designing a neural network architecture, different free parameters have to be decided. Among these parameters are various numbers of layers, mainly numbers of hidden layers, numbers of neurons in hidden layers, numbers of training iterations, etc. Input layer does not have free parameters, but it is largely dedicated by the number of input receivers. Output layer also has no free parameters, as it defines the number of classes. In hidden layer, each node and layer has free parameters as have many number of hidden layers as well as neurons in it. The number of hidden neurons, learning rate, and momentum parameter as well as output from each neuron, which should be more than the threshold value, are very important parameters in terms of training of model and network design of model. The hidden layers and their nodes contribute decision boundaries for discriminating one class from the rest. Unlike the conventional statistical methods, these learning methods do not define the decision boundaries of each class by their statistical distribution in feature space. Hence ANN is allowed to self-learn its the decision boundaries. The exact required number of hidden layers and neurons in hidden layers as well as initial weights are not known a priori.

A general approach is to start with a large number of hidden layer neurons and trim or "prune" the network to optimal size through a strategy to evaluate the impact of each neuron elimination in the final output evaluation. In other words, not all hidden neurons contribute significantly to the output performance and hence can be "pruned" out.

In hidden layers, only hidden neurons as defining hidden layer size are present. Neurons in hidden layers are essentially hidden from view due to being bonded between the input and output layers. The size of hidden layers, called number of neurons in hidden layers, is generally free of parameters in the ANN system. More the hidden layers as well as hidden neurons in them require greater the processing power and system flexibility. This additional flexibility can be attained with the cost of added intricacy in the ANN training algorithm. If the ANN gets over-specified and is in general incapable of generalization while having large numbers of hidden neurons. It is similar to a system of mathematical functions with more free variables. The effect of few hidden neurons, equally, will avert the system in appropriately fitting the input data and result in less robustness of the system.

Hidden layer size is defined as the number of neurons in the hidden layers, which are the layers between input and output layers. According to Huang (2003), the number of hidden nodes in two-hidden-layer case are always more than the output layer neurons. Lippmann (1987) and Pao (1989) also demonstrated how neural networks can be used as supervised and unsupervised modes. One of the parameters in neural networks which is very common is learning rate. Learning rate controls the change in weight and bias changes while training algorithm is applied. Another parameter of learning is momentum adding a fraction "*m*" from previous weight to update the current one. The momentum parameter prevents the ANN from ending on a local minimum or saddle point. But momentum increases the speed along a local minimum as well as along flat regions. When considering a large value of a momentum parameter, momentum increases the speed toward convergence. However, applying too high a value of momentum parameter can generate the risk of jumping over the global minimum, and the system becomes unstable due to overshooting. A momentum coefficient with values that are too low cannot consistently avoid local minima, and learning algorithms training will slow down. Another parameter of ANN is epochs, which controls learning. Epochs determine when training will stop, based on whether the number of iterations exceed the epochs. Secondly training can stop when the training neural network reaches the minimum error criteria, or reaches to maximum number of iterations. The minimum mean square error of the epoch is used to monitor learning cycles. Mean square error is calculated using square root of the sum of squared differences between the network targets and actual outputs divided by number of patterns.

Training of ANN has to be done with training data before using ANN for classification or prediction. The question becomes, how does the learning process take place in ANN? Inputs as training data are used in the learning process, and the outputs are represented in numeric form. In the training process, weights initially assigned randomly as internal state of ANN are modified. Output calculated using tuned weights is as close as possible to the desired output but in the predicted process generates most likely output according to its past "training experience" – using the weight values as internal state, based on input. With this, machine learning is also called the model fitting approach. To understand this learning process, it can be decomposed into its several building blocks.

To understand supervised learning of ANN, here a single dimension vector has been taken. In this, there will be input as a single value and output single dimension to understand the relation between input and output. The objective of the supervised neural network as a best fit model is to find out the characteristic function that fits the best for a given data. The initial hypothesis in supervised classification is learning the model where training data is used to adjust it's weight values. Neural networks can start while taking random weights value at each connectivity. At each connectivity there should be weights, but initially at each connectivity there are no weights. Thus, the simplest procedure is to initialize weights randomly, called learning model initialization.

In order to make the ANN model to generalize to any problem, there is a function called loss function. How ANN reaches to its goal of generating appropriate outputs as close as possible to the desired values can be known using performance metrics generated from the loss function.

The simplest intuitive loss function is simply loss = (Desired output – actual output). This loss function returns positive values when the network undershoots (prediction < desired output) and negative values when the network overshoots (prediction > desired output). To generate the loss function, an absolute error on the performance regardless if its overshooting or undershooting can be defined as: loss = Absolute value of (desired – actual).

Different errors from various situations can give same sum of errors; for example, a large number of small errors or very few big errors can accumulate exactly to the same total amount of error. It has been more preferable to have a distribution of many small errors, rather than a few big ones, while to work under any situation. ANN can converge to a situation while defining the loss function to be the sum of squares of the absolute errors. These types of loss functions are the most famous loss functions in ANN. The advantage of considering small errors is that they are counted much less than large errors.

Notice how, for first input, the network predicts correctly the result. What one has to care about is to minimize the overall error from the whole dataset that is the total of the sum of the squares of the errors. Overall loss function is an error metric that gives an indicator on how much precision is lost, while replacing the real desired output by the actual output generated by the trained neural network model, which is called loss.

In machine learning, the objective is to *minimize* the loss function and bringing loss value as close to 0 as possible. During transform of machine learning, a problem can be solved using an optimization process that aims to minimize loss function. Obviously, an optimization technique is what modifies the weights of neural networks, so that total function loss can be minimized. Optimization can be done through genetic algorithms, greedy search, or even a simple brute-force search. In the case of only one parameter of weight to optimize W, the search can be from $-\infty$ to $+\infty$ with a small rate change to find out at which weight value the sum of squares of errors is smallest over the dataset.

This model works fine even though the model has only very few parameters and not much concern about precision. However, like in image processing, if training

ANN over an array of 600 × 400 inputs, weights will be in the millions to optimize. This means that as dimensionality of data increases, number of neurons increases in input layer, due to which the number of weights increases. Fortunately, there are various methods in mathematics which can guide how to optimize the weights in ANN, and one of the methods is differentiation. This differentiation is generated with the help of derivative of the loss function. The derivative of a function in mathematics controls the rate by which a function is changing its values at a given time. Read Bishop (1995), Garson (1998), Haykin (1999), Hewitson and Crane (1994), and Ripley (1996) to know more about working with artificial neural networks.

4.2.1 Back-Propagation

In neural networks between input and output layers, zero or more than zero hidden layers can be used inside the neural network. The effect of a large number of hidden layers is to have more variations in the working of the neural network, which can be achieved (Kavzoglu, 2001). Kanellopoulos and Wilkinson (1997) proposed that when there are 20 or more classes, two hidden layers should be considered, with neurons in the second hidden layer two to three times more than the output layer. Sarle (2000) mentions reasons to highlight the causes for determining number of neurons in the hidden layer. The reasons are on total input and output neurons present, training data set size, classification complexity, noise in the input data to be classified, hidden layer neuron's activation function, and type of training algorithm.

Training of ANN means using some training algorithm to adjust the weights present at each connection of ANN (Figure 4.2). In order to adjust the weights during training, derivatives take care about change in weights through the back-propagation algorithm. First stage error is received from ANN through loss function, with its derivate, and with the derivation of each function from the composition, it is possible to propagate back the error from the end to the start. To implement feed-forward, there can be a set of libraries of differentiable functions, by directly applying the function. Secondly, through these libraries backpropagation can be applied with the derivative of the function, to combine a complex

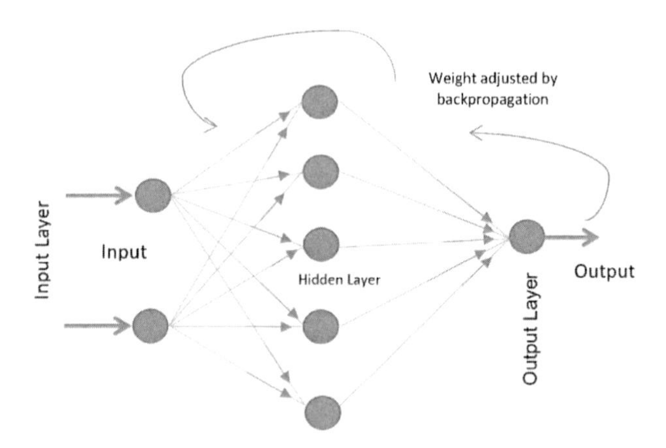

FIGURE 4.2 Back-propagation for adjusting weights.

neural network. During the forward pass their parameters, also called the track of the function calls, have to be kept. It is possible from derivatives of these functions to know the back-propagated errors. This is achieved by de-stacking of the function calls. This is called the auto-differentiation approach, and requires only the function which is provided with the implementation of its derivative. Through implementing basic mathematical operations over matrices, auto-differentiation can be accelerated. In neural networks, the previous layer neurons forward their output to the next layer neurons and finally to neurons of output layers, in order to implement back-propagation for training of ANN. In back-propagation, error at the output layer is used to adjust the weights in the previous layer, later output coming out from the previous layer, then the output layer is used to find out error and further adjust the weights from its previous layer. Figure 4.2 shows the process of implementation of back-propagating errors. At each stage of ANN, such as Input → Forward calls → Loss function → derivative → back-propagation of errors, there is correction factor delta on each weight of the training stage.

4.2.2 Weight Update

While adjusting weights in training of ANN, the error change rate is represented as the derivative (Figure 4.3). It has been observed that, in real data processing, weights cannot be modified with very higher rates. Kavzoglu (2001) mentions that learning rate value 0.2, with momentum used; with momentum it should be 0.1 to 0.2. Pal and Mitra (1992) suggested parameters be changed during training. As the data have lots of non-linearities, any very large change in weights may not able to incorporate adjustment as required. It has been found that derivative is only local at the area. Weight updates general rule in ANN can be called the delta rule:

$$\textit{New weight} = \textit{Old weight} - \textit{Derivative Rate} \times \textit{Learning rate}$$

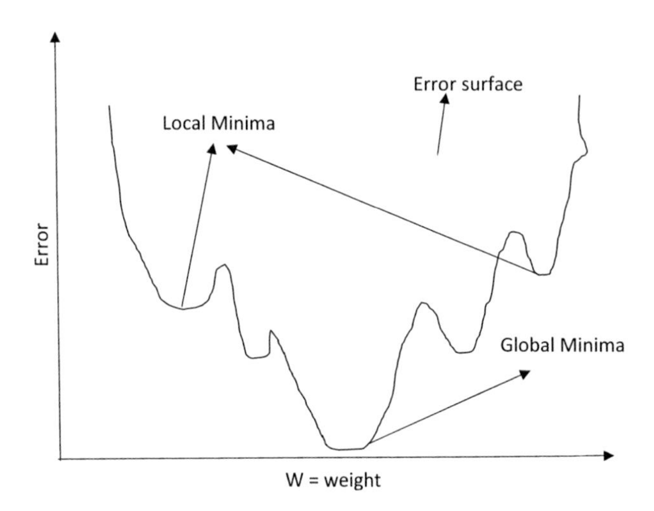

FIGURE 4.3 Local to global minima during training.

The learning rate is constant as used in ANN; this has a small value to update weight values smoothly and slowly and to avoid big stepladders and messy behavior. To validate this concept, the following points can be considered:

- When derivative rates are positive, there is an increase in weight, which will increase the error, so new weight should be smaller.
- When derivative rates are negative, there is an increase in weight, which will decrease the error, thus weights should be increased.
- When derivative is 0, it is in stable minimum. When it reaches the stable state, no updates are required.

In the present scenario, several weight update methods used in ANN are often called optimizers. The delta rule as weight optimizer is the most simple and intuitive one; however, it has several drawbacks. In the testing, sample data input/output training set may be small, but in the real world data size may be very large. While considering batch learning, error cost function or loss function is minimized, considering the whole dataset. But this, called batch learning, might be very slow for a large dataset. To increase the speed, sub-batches can be made, providing that the dataset is shuffled randomly, which is called mini-batch gradient descend (Benediktsson et al., 1993). If considering each single input, output observation in which weights are updated can be called full online learning or stochastic gradient descent. So, any optimizer can work with three optimizer modes like full online/mini-batch/full-batch. Kavzoglu (2001) found classifier performance better with small initial weight in the range of [−0.25, 0.25].

While weight updated rate with a small delta step, the number of iterations will increase in order to learn. The same thing happens in genetic algorithms in which with each generation, the fittest survives after a small mutation rate is applied. In neural networks, after each iteration, the gradient descent force updates the weights in such a way that global loss function is minimized. The similarity between ANN learning and genetic algorithms can be seen through the delta rule, which acts as a mutation operator in genetic algorithm and the loss function acts as a fitness function to minimize in ANN. Mutation in genetic algorithms is blind – that is the difference. Good mutations have a higher chance to survive. The weight updates in ANN are controlled through decreasing gradient force over the error. The number of iterations required to converge may depend on the following points:

- Effect of learning rates is like large learning rate with higher chance of instability.
- Effect of learning also depends on free parameters of the network like how many layers, activation function, etc. The larger the number of variables, the more time to converge, but the higher the accuracy ANN reaches.
- There are also effects of learning algorithms used; some weight adjustment rules are proven to be better than others.
- There is also the effect of weight initialization of the network. Weights are initialized in such a way that there is only step away from the optimal solution.
- Effect of quality of the training set also strongly affects learning of ANN.

4.3 CONVOLUTIONAL NEURAL NETWORK (CNN)

In various applications such as self-driven cars, smart web searches, and pattern of speech and image recognition the application of machine learning has grown many fold in last decades (Bhandare et al., 2016). Machine learning applications are becoming part of daily life (LeCun et al., 2015). In machine learning, a branch of artificial intelligence computers teaches itself with being provided reference or training data. Machine learning could drive future technology, being a very interesting and complex topic (Krizhevsky et al., 2012; Long et al., 2015; Dong et al., 2016).

The machine learning algorithm follows the principle of biological neural networks the same as biological neurons are organized. Machine learning gives an opportunity to mimic the processes arising in the brain (Zeng et al., 2015). One of the examples of machine learning is a neural network; the neural network consisting of individual units called neurons. Neurons are located in all the layers of a neural network. Neurons in each layer are connected to neurons of the next layer, and data flows from the input layer to the hidden layer to the output layer. Output from each neuron is generated using a weighted sum into an activation function present in each neuron. Output from each neuron is transmitted to all the neurons of the next layer connected to it.

The benefit of neural networks came in connection with high end processing machines and large training data. With these two points, modified versions of neural networks can be seen used in last decade, including convolutional neural network (CNN), recurrent neural network (RNN), etc., added on with deep learning concepts. In these networks, technical pre-processing stage structures of neural networks are different than ANN. But these modified networks are able to solve a wide range of tasks which were not effectively solved in the past. From various applications of these networks, image classification can be an important example.

4.3.1 Convolutional Neural Networks Image Classification

A convolutional neural network (CNN) has different pre-processing layers in comparison to MLP neural networks as shown in Figure 4.4. Basically, CNN uses features called visual cortex (Hubel and Wiesel, 1968; Fukushima, 1980). The visual cortex in the brain is the main cortical region. Its main work is to receive, integrate, and process visual information, which is received from the retinas. Due to mimicking like brain, the most popular use of CNN architecture is image pattern recognition. Some of the applications of CNN in social media are automatic tagging in Facebook photographs, in Amazon for generating product recommendations, and in Google for searching through photos.

In remote sensing for image data, the main use of CNN is for earth object recognition. The main objective of object recognition is identifying similar patterns in one group and providing labels. The pattern recognition is a skill learned by people from birth, and they are easily able to determine various objects. But the way a computer identifies the objects from an image is quite different. The computer considers an image as an array of pixels. For example, an image can be of size 300×300 with RGB bands. In this example, the array size will be of $300 \times 300 \times 3$,

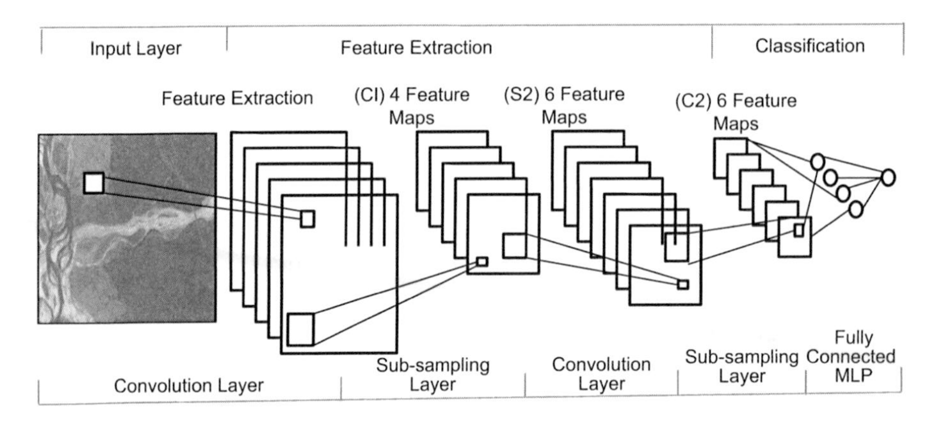

FIGURE 4.4 CNN – Multi-layer CNN architecture.

where 300 are rows, the next 300 columns, and 3 is RGB channel values. In the image range of pixels, value depends on data types of the image, which gives bits of the image. For 8 bits, data images have values from 0 to 255 to each of these numbers. These values describe as vector element the intensity of the pixel at each row and column. To classify these pixels, a unique property is generated from a classification algorithm. In human learning, these characteristics can be, for example, specific characteristics of an object. For the computer, these characteristics are different shapes of an object. In the case of convolutional neural networks, convolutional layers construct more abstract concepts. The steps followed in CNN on a given input data are a series of convolutional operations, non-linear operations, pooling layers operations, and last, fully connected layer steps are applied to get the output.

The Convolution layer operation is always the first in which an image is input to it. Reading of an image is done from the top left corner of an image, as this takes the least time in reading an image. The next step is to select a small matrix, called a filter. This filter generates convolutions output, while moving along the input image. This filter job is to multiply its values with the original pixel values of the image in that region. Multiplying filter coefficients with image values weighted sum is calculated, and finally a single number is generated from this operation. Initially the filter has used pixel values from the upper left corner only, and it moves further and further right by 1 or higher unit, performing a similar operation. This unit movement is called stride, and stride can be of any unit. After applying the filter across the image, an output in the form of a matrix is obtained, which is smaller than an input matrix. This operation identifies boundaries and simple colors from input image with respect to human perception. In order to recognize very specific properties of features, a higher level of CNN network is required. The CNN network consists of several convolutional operations, pooling operations, and non-linear layers. In CNN, the first layer's output becomes input to the next layer, even though the next layer may not be the same as the first one, and this happens to consecutive layers until it reaches the final layer.

In CNN activation, a function is present as non-linear layer, after each convolution operation to be applied (Glorot et al., 2011; Krizhevsky et al., 2012; Nair and Hinton, 2010; LeCun et al., 2015; Ramachandran et al., 2017). This function produces a non-linear decision boundary via non-linear combinations of the weight and inputs. Without the non-linear decision boundary, the network would not be able to model the response variable. After the non-linear layer, the pooling layer comes in CNN (Lin et al., 2013). Pooling operation reduces the image size through down sampling operation. In down sampling, the image is compressed in which its details were reduced as less detailed pictures. This down sampling operation was ended as various features have already been identified in previous convolution operation, and an image with detailed information is not needed.

In CNN, after various steps of convolutional operations, non-linear operations and pooling layers are applied, then the last compulsory operation layer, called fully connected layer, has to apply. Fully connected layer gives the final output information from convolutional networks. The end layer in CNN, called the fully connected layer, provides N dimensional vector; N depicts the number of classes to identify.

4.3.2 Supervised Machine Learning

In supervised machine learning, the model gets trained using training data and expected output data generated from input data. While creating CNN types of models, some of the steps have to follow, such as model structure creation, model weight adjustment through training, and model evaluation through testing process. The machine learning algorithm decides model construction. Input data should be scaled (to 0 to 1 range) before model training. Once a model has been constructed, learning of the model has to be done. Learning of the model is conducted using training data and evaluated expected output for this data with target data. Model testing is done after the model training is completed. The second set of data is loaded to test the model. Testing data was not used with models previously to have unbiased accuracy of the model (Park and Han, 2018). Once model training is over and the model gives correct results, its weight values and other parameters have to be saved. The model with saved parameters and weights can be used with real data sets; this process can be called model evaluation. From this evaluation, the process model can be used to evaluate new datasets. The pseudo code of CNN with an option to add LSTM layers is given as follows:

```
class CNN:

    def __init__(self,bands,train_cycle,c_c,thresh,D_
type,row,col):
            self.bands=bands
            self.row=row
            self.col=col
            self.pixels=row*col
            self.train_cycle=train_cycle
            self.c_c=c_c
            self.thresh=thresh
```

```
            self.D_type=D_type
            print(self.pixels)

    def cnnfit(self,data,X,y_train,activation_fn):
                print("X")
                print(X.shape)

                print("Y")
                print(y_train.shape)


                print(self.c_c)
                model = Sequential()
                j=3
                for i in range(0,len(data)):
                    print(data[i])
                    if data[i]=="Conv":
                        if(i==0):
                            print("yes",self.bands)
                  model.add(Conv1D(2 ** j, 2,
activation=activation_fn, padding='same',
input_shape=[self.bands, 1]))
                        else:
                           model.add(Conv1D(2 ** j, 2,
activation=activation_fn, padding='same'))
                        #if i!=len(data)-1:
                            #model.add(BatchNormalization())
                        j=j+1
                    elif data[i]=="Maxpooling":
                        model.add(MaxPooling1D(2))

                    elif data[i]=="LSTM":
                        if(i==0):
                      model.add(LSTM(2**j,return_
sequences=False, input_shape=(bands ,1)))
                        else:
                           model.add(LSTM(2**j, return_
sequences=True))
                        j=j+1

                model.add(Flatten())
                model.add(Dropout(0.1))
                model.add(Dense(256, activation= activation_fn))
                model.add(Dense(128, activation=activation_fn))
                model.add(Dense(self.c_c, activation='softmax'))
                model. compile(loss='categorical_crossentropy',
optimizer='adam', metrics=['accuracy'])
```

The following code snippets are the keras implementation of convolution neural networks. The network consists of a set of three layers in the order: 2D convolution, non-linear activation layer, rectified linear unit (ReLu), and pooling layer (max pooling or average pooling). The output of the final layer acts as an input to the following two dense layers. Below is the more detailed structure of the model.

 # MODEL

Algorithm 1: Pseudo code for CNN Model

Requirement: Training Dataset: features *x_train*, labels *y_train*
Testing Dataset: features *x_test*, labels *y_test*

Procedure:

```
NeuralModel(x_train, y_train)

batchSize =  16; epochs = 10

model = Sequential()

model.add(Conv2D(32, (3, 3), input_shape=(300, 300, 3)))
model.add(Activation('relu'))
model.add(MaxPooling2D(pool_size=(2, 2)))

model.add(Conv2D(32, (3, 3)))
model.add(Activation('relu'))
model.add(MaxPooling2D(pool_size=(2, 2)))

model.add(Conv2D(64, (3, 3)))
model.add(Activation('relu'))
model.add(MaxPooling2D(pool_size=(2, 2)))

# the model so far outputs 3D feature maps (height,
width, features)

model.add(Flatten())  # this converts our 3D feature maps to
1D feature vectors
model.add(Dense(64))
model.add(Activation('relu'))
model.add(Dropout(0.5))
model.add(Dense(1))
model.add(Activation('sigmoid'))
# COMPILE
model.compile(loss='binary_crossentropy',
       optimizer='rmsprop',
       metrics=['accuracy'])

model.fit(x_train, y_train, batch_size=batchSize,
epochs=epoch)
score = model.evaluate(x_test, y_test, batch_size=32)
accuracy = score[1]
return accuracy
```

end procedure

Now let us understand the individual layers in more detail. The first convolution layer Conv 2D takes three input arguments in the following order: number of output filters, a tuple of kernel size which determines the dimensions of convolution window, and a tuple of input shape. As mentioned in the code, there are 32 output filters each of dimension (3, 3) in the first layer. It is worth noting that only first convolution layer has a parameter of input dimension (300, 300, 3), while the subsequent layers automatically take in the value of input dimension from the previous layers. Thus all the convolution layers except the first take two input arguments.

The model uses ReLu as an activation function. The function provides a non-linearity to the input values as $f(x) = \max(0, x)$. The pooling functions (Max pooling, Average pooling) are applied on the spatial data, and the number (2, 2) means that the output's dimension will be half of the input dimension based on the type of pooling performed.

Followed by the convolution layers are two dense layers which have ReLu and Sigmoid as activation functions. The first dense layer has 64 neurons, the outputs of which are fed to the dropout layer followed by a single layer of a neuron with a sigmoidal function as the activation. In the model, the dropout plays an important role as it helps us to cope with the issue of overfitting. Overfitting refers to a phenomenon in which the model performs extremely well on training data while giving very low accuracy on unseen test data, the reason being that the model has been tuned to training examples so well that it fails to classify any new unseen example. Thus, the dropout function 'drops' in some neurons, making the model more general than being too specific based on the examples. Once the model has been built, it is compiled using "binary_crossentropy" as loss, rmsprop as the optimizer, and "accuracy" as the metric to access the model performance quantitatively. The following code fragments help us construct the model for training:

```
batch_size = 16
```

4.4 RECURRENT NEURAL NETWORK (RNN)

Sequence prediction problems have been used to design recurrent neural networks or RNNs (Figure 4.5). Depending upon types of inputs and outputs in an application area, various types of sequence prediction problems can be considered (Robinson et al., 2002; Graves et al., 2013; Sak et al., 2014). Various types of sequence prediction problems can be: One-to-Many: Multiple steps as an output can get through mapping with input. Many-to-One: Class or quantity prediction gets mapped with sequence of multiple steps as input. Many-to-Many: Multiple steps as output mapped with sequence of multiple steps of input. Sequence-to-Sequence or seq2seq in short is a many-to-many problem. Training of recurrent neural networks is a difficult task. To solve the training problem of RNN, the best effective network is long short-term memory, or LSTM (Hochreiter and Schmidhuber, 1997; Mikolov et al., 2010; Sundermeyer et al., 2012). The advantage of use of LSTM resolves the issues of training a recurrent network, and the LSTM has been applied on a wide range of applications. RNNs as the LSTMs have received the most success while applying in sequences of words and paragraphs, in the area of natural language processing (Elman, 1990). Recurrent neural networks (RNNs) are networks in which data flows in different directions

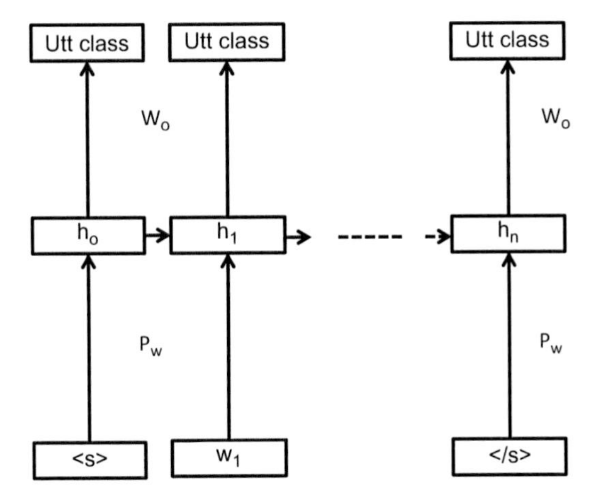

FIGURE 4.5 RNN architecture with different layers.

and are used for applications such as language processing (Chung et al., 2014). For RNN, use of long short-term memory is effective. Time series data is represented as an example like a series of text and spoken language (Zhao et al., 2016; Wang et al., 2017). Generative models have also been used that require a series of output, not only in text form, but also in creating handwriting. RNNs have been applied for speech data, regression prediction problems, generative models application areas, text data, and classification prediction problems (Mandic and Chambers, 2001; Sutskever et al., 2011). For tabular datasets, recurrent neural networks are not appropriate, and RNNs are also not appropriate for image data input. Time series forecasting problems have also been tested with RNNs and LSTMs, but the results have been poor (Walid and Alamsyah, 2017). Simple MLPs outperform LSTMs when applied on the same data. In computer vision, a recurrent neural network (RNNs) has great application. In acoustic modeling for automatic speech recognition (ASR), RNNs also have been strongly applied (Lekshmi and Elizabeth, 2016).

4.5 HYBRID LEARNING NETWORK (HLN)

From literature, hybrid network models have been found while adding in one model layers of other models like, CNN layers with the RNN model (Guo and Ding, 2015). These types of HLN networks in a broader model have different combinations of MLP layers. Technically, HLN having different layers of ANN, CNN, and RNN can be called hybrid type network architecture (Chin-Yi and Chung-Wei, 2012). Adding different types of network layers, known as 'mixing' of layers, together make hybrid models. Hybrid models will have a combination of layers mounded one at the top such as CNN layer, pooling layer, LSTM layer, and MLP or LSTM at the output (Chiou-Jye and Ping-Huan, 2018). CNN LSTM architecture types of models work with series of images like video frames and generate classification or prediction outputs.

To unlock new capabilities of networks, these models can be stacked with specific layered architectures to use in reusable image recognition prototypes. Hybrid models use a large range of layers as deep CNN and MLP networks with LSTM model and can be applied for captioning photos (Hu et al., 2018). Input and output sequences of differing lengths can be archived through encoder–decoder LSTM networks. The project outcome must be very clear and then seek out a network architecture that meets specific project needs.

4.5.1 Training Issues – Remote Sensing Data Domain

It is well known that learning algorithms require large training data sets. But while applying these learning algorithms on remote sensing data sets, it becomes impossible to collect large training data sets. To resolve this issue, a technique has been developed to increase the training data size. Like in photogrammetry, the extension of ground control points (GCPs) approach is there. In the case of extending/increasing the number of GCPs from a few number, it is achieved by triangulation technique. In the same manner, the training data size for learning algorithms can be increased from small seed training to larger data size, using statistical parameters.

4.6 DEEP LEARNING CONCEPTS

A deep neural network (DNN) basically has more layers in comparison to a basic neural network (ANN), between input and output layers (Vargas et al., 2017). But the main point in DNN is generating features at very large dimensionality so that some of the information which cannot be identified in normal features of data can be identified in the high dimensionality feature of data sets. Larger dimensionality of data can allow seeing specific properties in data to identify specific object (Bengio, 2009). The DNN generates better mathematical processing to convert the input into the output for a linear or non-linear relationship (Alom et al., 2019). The probability of each output is calculated due to operations done across layers. For example, degree of belongingness of a tree in a certain type of forest can be identified through a trained DNN. In network based algorithms, in each layer mathematical manipulations take place, and complex DNNs have large number of layers and hence become "deep." They are also called deep due to generating high dimensional data from a given input image. DNNs are capable of handling complex non-linear relationships. Objects can be expressed as a layered composition of primitives if DNN architectures generate compositional models. Extra layers in DNN enable arrangement of features from fewer layers like in ANN. These extra layers model complex data in comparison to similarly performing fewer layers in ANN, where each layer is called a unit.

Deep architectures in deep learning have different variants of layers. Specific architecture of layers has given good results in specific applications. Comparison of multiple layers structure with DNN cannot be done unless both performances are evaluated with the same data sets. Same as ANN, DNNs are classically feed-forward networks, which means in DNN also, without looping back, data flows from the input layer to the output layer. Like in ANN, in DNN randomly numerical weights also are assigned in fully connected layers. The weights are considered just like in ANN where weighted sum is calculated using weights and inputs. This weighted sum goes inside the neuron and returns an output between 0 and 1. While working with DNN, some of the input patterns are not recognized correctly; in that case, a learning algorithm would adjust the weights. While identifying unknown data during classification, there can be certain parameters that are more influential; these parameters have to be tuned until correct values of these parameters is determined.

4.6.1 Challenges in Learning Algorithms

Similar to ANNs, in DNNs many issues can arise. Overfitting and computation time are the common issues in CNN, RNN, DNN, etc. (Hinton et al., 2012). Due to added layers of abstraction, DNNs are prone to overfitting, which can be resolved through modeling enslavements in the training data. To resolve overfitting, regularization methods such as Ivakhnenko's unit pruning or weight decay (regularization) or sparsity (regularization) can be used during training. Also, training data size can be increased. From the hidden layers during training, dropout regularization randomly omits units, which helps to exclude rare dependencies. Overfitting can be reduced if training data can be amplified using approaches like shifting, flipping, cropping, and rotating so that training data size can be enlarged (Ioffe and Szegedy, 2015; Zhong et al., 2017).

There are large numbers of free parameters in DNNs such as types of layers, its units or filters, learning rate, momentum, and initial weights. To reduce large time and processing resources, sweeping in the possible parameter values to achieve optimized parameters may not be suitable. Speed of computation can be improved through various tricks, such as dividing tasks into batches such as computation of gradient on small sets training samples at a time, rather than whole training data (Qian, 1999; Kingma and Ba, 2014; Ruder, 2016). Because of the availability large core architectures like GPUs or Intel Xeon Phi, matrix, and vector, computations have produced significant speedups in training.

With more straightforward and convergent training algorithms, type of neural networks can be explored. One such kind of neural network is called CMAC – cerebellar model articulation controller. In CMAC, learning rates or randomly generated initial weights are not required. The training activity while applied in a batch can be guaranteed to converge, while arithmetical computation in training algorithm behaves linearly, dependent on number of neurons considered.

One of the first convincing successful uses of deep learning is in automatic speech recognition. Intervals of multi-second containing speech events tagged by thousands of discrete time steps, with one time step equivalent to about 10 ms, can be applied in LSTM RNNs to have very deep learning. With traditional speech recognizers on certain tasks, LSTM with forget gates is competitive (Han et al., 2017).

Deep learning has been related to theory of working of brain proposed by cognitive neuroscientists in the early 1990s. Computational models have been made by predecessors of deep learning systems following instantiated developmental theories. Developmental models generate characteristics with various important learning dynamics in the brain, supporting self-learning. These are somewhat similar as in ANN applied in deep learning structures. In DNN, having various layers with different operations on the input data generates information from a prior layer and then passes its processed output to the next layers.

To explore the likelihood of deep learning models from neurobiological perspective, variant methods are developed. In order to increase backpropagation algorithm's processing realism, several variants have been proposed. Some researchers have identified hierarchical generative models and deep belief networks as unsupervised approaches of deep learning which may replicate biological reality. In sampling based processing in the cerebral cortex, generative neural network models are similar to neurobiological evidence.

A systematic study between the human brain and deep neuronal networks is required to find out comparisons between both. Still, for example, in the context of computations, performance between deep learning units and actual neurons and neural populations could be similar. Similarly, deep learning models are similar to those cardinal visual systems, which contain dozens of distinct areas in the cerebral cortex that are widely interconnected in an isolated hierarchical network that contains several entangled processing streams.

4.7 IN-HOUSE TOOL FOR STUDY OF LEARNING ALGORITHMS

Pseudo code has been implemented in following GUI (Figure 4.6). This GUI can be used for displaying multi-spectral remote sensing data. This GUI has facility to collect reference data of various classes to be used for training and testing purposes (Figure 4.7). Reference data of each class is saved in different data files. In the next step, this GUI facilitate to read and insert training data of each class (Figure 4.8).

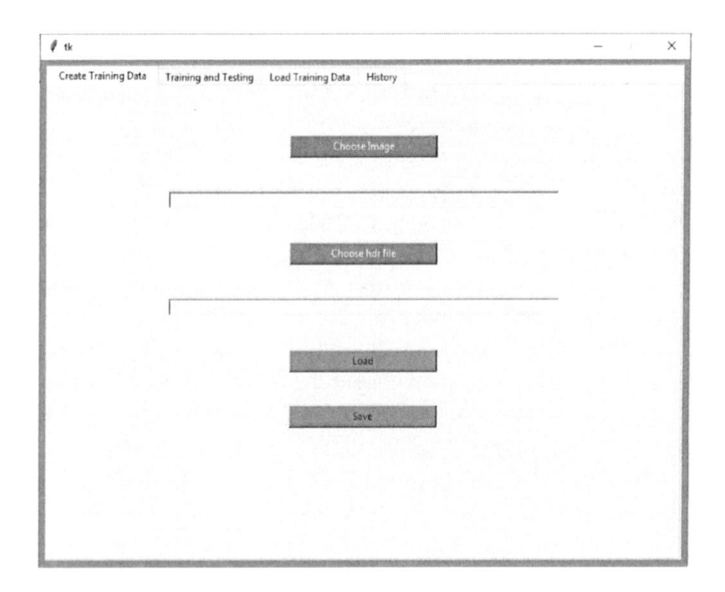

FIGURE 4.6 Main GUI of learning tool.

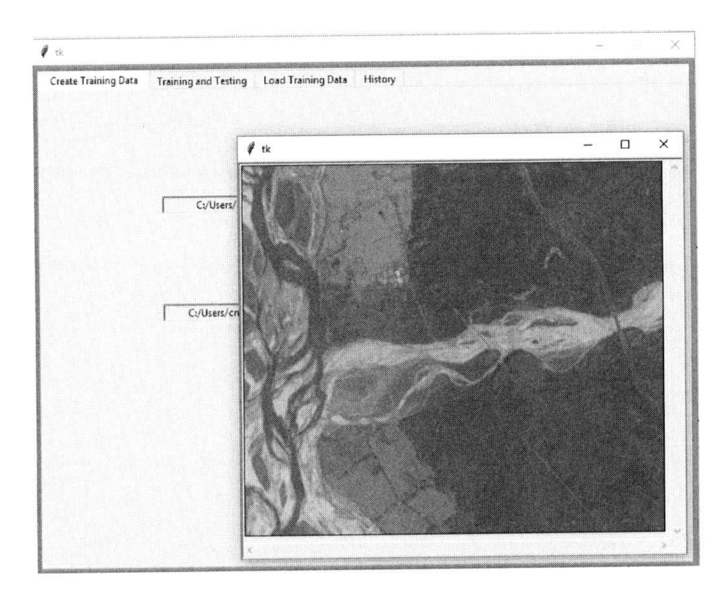

FIGURE 4.7 Training data creation tool.

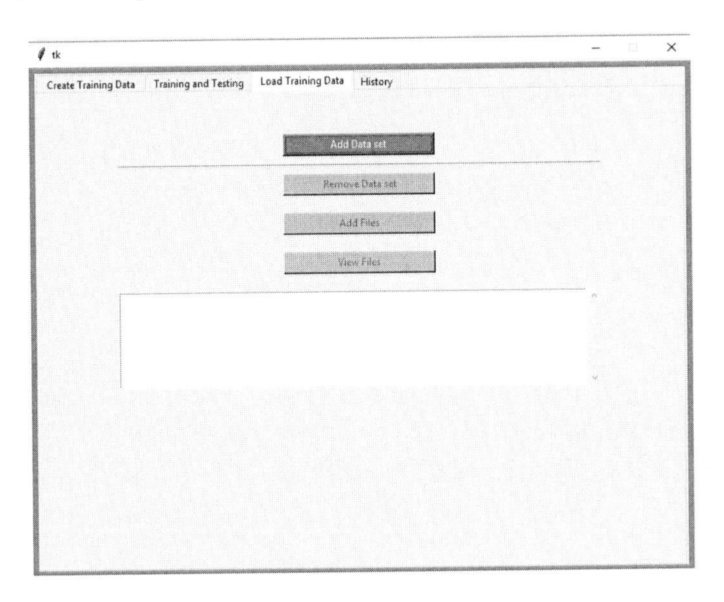

FIGURE 4.8 Load training data module.

Once training data of each class has been loaded, the next step comes to apply ANN or CNN networks, through training and tested module. By default it gives an option to apply ANN with input images, called training data, defining ANN parameters like learning rate, epoch, and hidden neurons (Figure 4.9). While going to CNN module, this parameter of CNN has to be defined like epoch or threshold (Figure 4.10). CNN module provides options to add CNN based layers as well as to make it a hybrid

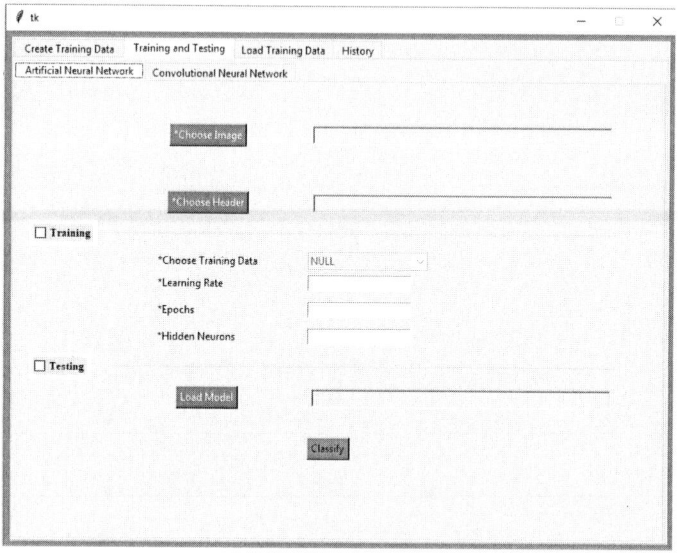

tk
Create Training Data | Training and Testing | Load Training Data | History
Artificial Neural Network | Convolutional Neural Network
*Choose Image
*Choose Header
Training
*Choose Training Data NULL
*Learning Rate
*Epochs
*Hidden Neurons
Testing
Load Model
Classify

FIGURE 4.9 Training and testing module with ANN.

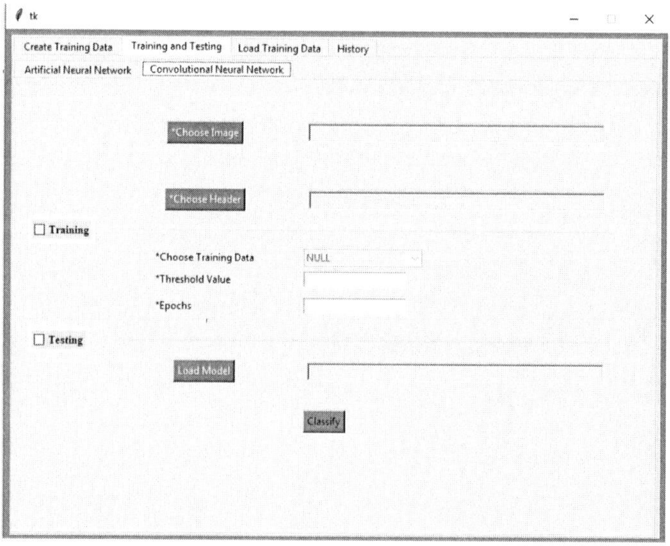

tk
Create Training Data | Training and Testing | Load Training Data | History
Artificial Neural Network | Convolutional Neural Network
*Choose Image
*Choose Header
Training
*Choose Training Data NULL
*Threshold Value
*Epochs
Testing
Load Model
Classify

FIGURE 4.10 Training and testing module with CNN.

approach while adding an LSTM layer (Figure 4.11). Once training is over, it will ask to save tunned weights of the model as well as hard and soft classified outputs. In hard classified output, it has been assumed that all pixels are pure pixels. In soft classified outputs, fraction images of each class is generated. During training and testing/classification, process accuracy and loss function status can also be monitored.

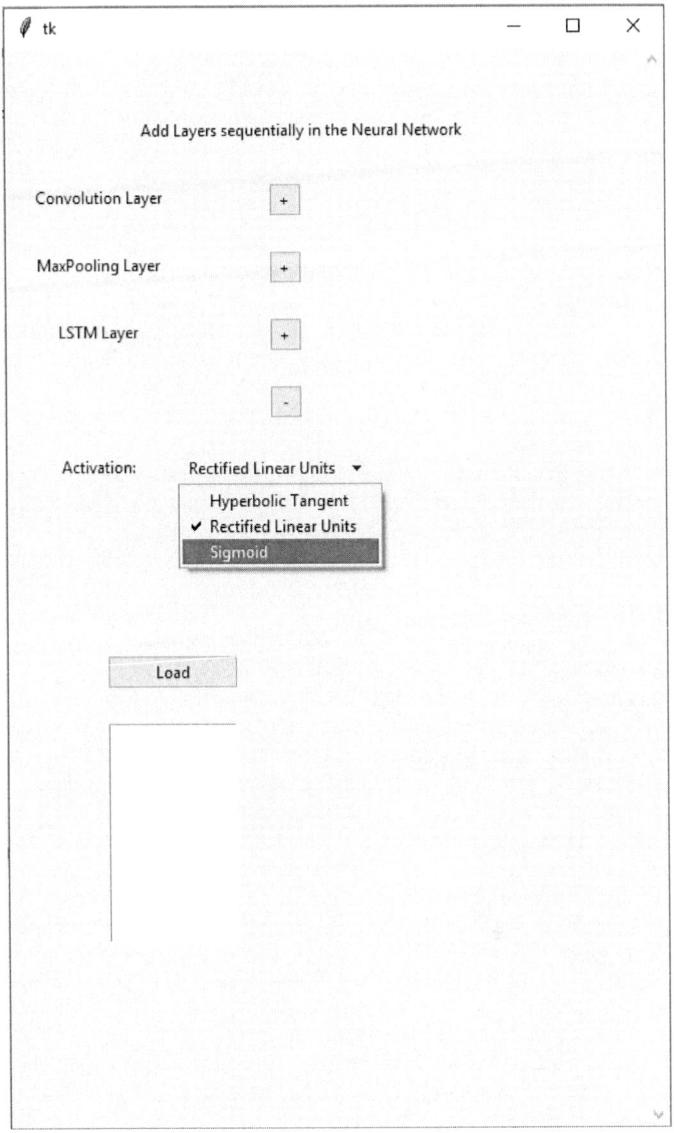

FIGURE 4.11 Module to add various layers in CNN module.

4.8 SUMMARY

In this chapter, starting from artificial neural network, convolutional neural network, recurrent neural network, hybrid approach, and deep learning concept have been covered. Information about each of these learning algorithms, their structure details, learning approaches, and advantages and disadvantages have been mentioned.

The main issue with these algorithms is limited training data availability, while applications on remote sensing data have been discussed and provided solution. The next chapter will be on understanding of hybrid fuzzy based classifiers.

BIBLIOGRAPHY

Alom, M.Z., Taha, T.M., Yakopcic, C., Westberg, S., Sidike, P., Nasrin, M.S., Hasan, M., Van Essen, B.C., Awwal, A.A. and Asari, V.K., 2019. A state-of-the-art survey on deep learning theory and architectures. *Electronics*, 8, 292. https://doi.org/10.3390/electronics8030292.

Benediktsson, J.A., Swain, P.H. and Esroy, O.K., 1993. Conjugate-gradient neural networks in classification of multisource and very-high-dimensional remote sensing data. *International Journal of Remote Sensing*, 14, 2883–2903.

Bengio, Y., 2009. Learning deep architectures for AI. *Foundations and Trends in Machine Learning*, 2, 1–127.

Bhandare, A., Bhide, M., Gokhale, P. and Chandavarkar, R., 2016. Applications of convolutional neural networks. *International Journal of Computer Science and Information Technologies* (IJCSIT), 7 (5), 2206–2215.

Bishop, C.M., 1995. *Neural Networks for Pattern Recognition*. New York: Oxford University Press.

Chen, C.-Y. and Li, C.-W., 2012. A hybrid network model to extract key criteria and its application for brand equity evaluation. *Mathematical Problems in Engineering*, 2012, Article ID 971303, 14. http://dx.doi.org/10.1155/2012/971303.

Chung, J., Gulcehre, C., Cho, K. and Bengio, Y., 2014. Empirical evaluation of gated recurrent neural networks on sequence modeling. In: *NIPS 2014 Workshop on Deep Learning and Representation Learning*. Montreal, Canada.

Dong, C., Loy, C., He, X. and Tang, K., 2016. Image super-resolution using deep convolutional networks. *IEEE Transaction Pattern Analysis and Machine Intelligence*, 38(2), 295–307.

Elman, J.L., 1990. Finding structure in time. *Cognitive Science*, 14 (2), 179–211.

Fukushima, K., 1980. Neocognitron: A self organizing neural network model for a mechanism of pattern recognition unaffected by shift in position. *Biological Cybernetics*, 36, 193–202.

Garson, G.D., 1998. *Neural Networks: An Introductory Guide for Social Scientists*. London: Sage Publications.

Glorot, X., Bordes, A. and Bengio, Y., 2011. Deep sparse rectifier neural networks. In: *Proceedings of the 14th International Conference on Artificial Intelligence and Statistics*, Fort Lauderdale, FL. vol. 15, pp. 315–323.

Graves, A., Jaitly, N. and Mohamed, A.R., 2013. Hybrid speech recognition with deep bidirectional LSTM. In: *Automatic Speech Recognition and Understanding (ASRU), 2013 IEEE Workshop on. IEEE*, pp. 273–278.

Guo, L. and Ding, S., 2015. A hybrid deep learning CNN-ELM model and its application in handwritten numeral recognition. *Journal of Computational Information Systems*, 11, 2673–2680. https://doi.org/10.12733/jcis13987.

Han, Z., Hong, M. and Wang, D., 2017. *Signal Processing and Networking for Big Data Applications*. Cambridge, UK: Cambridge University Press.

Haykin, S., 1999. *Neural Networks: A Comprehensive Foundation*. Englewood Cliffs, NJ: Prentice-Hall.

Hewitson, B.C. and Crane, R.G. (eds.), 1994. *Neural Nets: Applications in Geography*. London: Kluwer Academic Publishers.

Hinton, G.E., Srivastava, N., Krizhevsky, A., Sutskever, I. and Salakhutdinov, R.R., 2012. Improving neural networks by preventing co-adaptation of feature detectors. Available from https://arxiv.org/search/cs?searchtype=author&query=Hinton%2C+G+E.

Hochreiter, S. and Schmidhuber, J., 1997. Long short-term memory. *Neural Computation*, 9(8), 1735–1780.

Hu, Y., Wong, Y., Wei, W., Du, Y., Kankanhalli, M. and Geng, W., 2018. A novel attention-based hybrid CNN-RNN architecture for sEMG-based gesture recognition. *PLoS One*, 13 (10), e0206049. https://doi.org/10.1371/journal.pone.0206049.

Huang, C.-J. and Kuo, P.-H., 2018. A deep CNN-LSTM model for particulate matter (PM2.5) forecasting in smart cities. *Sensors*, 18, 2220.

Hubel, D.H. and Wiesel, T.N., 1968. Receptive fields and functional architecture of monkey striate cortex. *Journal of Physiology*, 195, 215–243.

Ioffe, S. and Szegedy, C., 2015. Batch normalization: Accelerating deep network training by reducing internal covariate shift. Available from https://arxiv.org/abs/1502.03167

Jeffrey, L.E., 1990. Finding structure in time. *Cognitive Science*, 14 (2), 179–211.

Kanellopoulos, I. and Wilkinson, G.G., 1997. Strategies and best practice for neural network image classification. *International Journal of Remote Sensing*, 18, 711–725.

Kavzoglu, T., 2001. An investigation of the design and use of feed-forward artificial neural networks in the classification of remotely sensed images. PhD Thesis, Nottingham, UK: School of Geography, The University of Nottingham.

Kenneth, C., Bruce, V., Kirk, S., John, F., Justin, K., Paul, K., Stephen, M., Stanley, P., David Maffitt, M.P., Lawrence, T. and Fred, P., 2013. The cancer imaging archive (TCIA): Maintaining and operating a public information repository. *Journal of Digital Imaging*, 26, 1045–1057.

Kingma, D.P. and Ba, J., 2014. Adam: A method for stochastic optimization. Available from https://arxiv.org/abs/1412.6980.

Krizhevsky, A., Sutskever, I. and Hinton, G., 2012. Imagenet classification with deep convolutional neural. In: *Advances in Neural Information Processing Systems 25*, Curran Associates Inc., Red Hook, NY, 1097–1105.

LeCun, Y., Bengio, Y. and Hinton, G., 2015. Deep learning. *Nature*, 521, 436–444.

Lee, D.H., 2013. Pseudo-label: The simple and efficient semi-supervised learning method for deep neural networks. In: *Proceedings of the ICML 2013 Workshop: Challenges in Representation Learning (WREPL)*, Atlanta, GA.

Lekshmi, K.R. and Elizabeth, S., 2016. Automatic speech recognition using different neural network architectures – A survey. *International Journal of Computer Science and Information Technologies*, 7 (6), 2422–2427.

Lin, M., Chen, Q. and Yan, S., 2013. Network in network. Available from https://arxiv.org/abs/1312.4400.

Lippmann, R.P., 1987. An introduction to computing with neural nets. *IEEEASSP Magazine*, April 4–22.

Long, J., Shelhamer, E. and Darrell, T., 2015. Fully convolutional networks for semantic segmentation. In: *Proceedings of the IEEE Conference on Computer Vision and Pattern Recognition (CVPR)*, Boston, MA, 3431–3440.

Mandic, D.P. and Chambers, J.A., 2001. *Recurrent Neural Networks for Prediction: Learning Algorithms, Architectures and Stability*. Hoboken, NJ: Wiley Online Library.

Mikolov, T., Karafiat, M., Burget, L., Cernocky, J. and Khudanpur, S., 2010. Recurrent neural network based language model. In: *Proceedings of INTERSPEECH, International Speech Communication Association*, Makuhari, Chiba, Japan, 1045–1048.

Nair, V. and Hinton, G.E., 2010. Rectified linear units improve restricted Boltzmann machines. In: *Proceedings of the 27th International Conference on Machine Learning*, Haifa, Israel.

Pal, S.K. and Mitra, S., 1992. Multilayer perceptron, fuzzy sets and classification. *IEEE Transactions on Neural Networks*, 3, 683–697.

Pao, Y.H., 1989. *Adaptive Pattern Recognition and Neural Networks*. Reading: Addison-Wesley.

Paola, J.D. and Schowengerdt, R.A., 1995a. A review and analysis of backpropagation neural networks for classification of remotely-sensed multi-spectral imagery. *International Journal of Remote Sensing*, 16, 3033–3058.

Paola, J.D. and Schowengerdt, R.A., 1995b. A detailed comparison of backpropagation neural network and maximum-likelihood classifiers for urban. *IEEE Transactions on Geoscience and Remote Sensing*, 33 (4), 981–996.

Park, S.H. and Han, K., 2018. Methodologic guide for evaluating clinical performance and effect of artificial intelligence technology for medical diagnosis and prediction. *Radiology*, 286, 800–809.

Qian, N., 1999. On the momentum term in gradient descent learning algorithms. *Neural Network*, 12, 145–151.

Ramachandran, P., Zoph, B. and Le, Q.V., 2017. Searching for activation functions. Available from https://arxiv.org/abs/1710.05941.

Ripley, B.D., 1996. *Pattern Recognition and Neural Networks*. Cambridge: Cambridge University Press.

Robinson, A.J., Cook, G.D., Ellis, D.P., Fosler-Lussier, E., Renals, S.J. and Williams, D.A., 2002. Connectionist speech recognition of broadcast news. *Speech Communication*, 37 (1–2), 27–45.

Ruder, S., 2016. An overview of gradient descent optimization algorithms. Available from https://arxiv.org/abs/1609.04747.

Sak, H., Senior, A. and Beaufays, F., 2014. Long short-term memory based recurrent neural network architectures for large vocabulary speech recognition. Available from https://arxiv.org/abs/1402.1128

Salimans, T., Goodfellow, I., Zaremba, W., Cheung, V., Radford, A. and Chen, X., 2016. Improved techniques for training GANs. Available from https://arxiv.org/pdf/1606.03498

Sarle, W.S., 2000. *Neural Network Frequently Asked Questions*. Available from ftp://ftp.sas.com/pub/neural/FAQ.html.

Stathakis, D., 2009. How many hidden layers and nodes? *International Journal of Remote Sensing*, 30 (8), 2133–2147, doi:10.1080/01431160802549278.

Sundermeyer, M., Schluter, R. and Ney, H., 2012. LSTM neural networks for language modeling. In: *INTERSPEECH*, Portland, OR, 194–197.

Sutskever, I., Martens, J. and Hinton, G.E., 2011. Generating text with recurrent neural networks. In: *Proceedings of the 28th International Conference on Machine Learning (ICML-11)*, Bellevue, WA, pp. 1017–1024.

Vargas, R., Mosavi, A. and Ruiz, R., 2017. Deep learning: A review. *Advances in Intelligent Systems and Computing*. https://eprints.qut.edu.au/127354/7/DEEP_LEARNING_A_REVIEW.pdf.

Walid, A., 2017. Recurrent neural network for forecasting time series with long memory pattern. *Journal of Physics: Conference Series*, 824, 012038. https://doi.org/10.1088/1742-6596/824/1/012038.

Wang, C., Wang, Y., Huang, P.-S., Mohamed, A., Zhou, D. and Deng, L., 2017. Sequence modeling via segmentations. In: *Proceedings of the 34th International Conference on Machine Learning*, Sydney, Australia, 3674–3683.

Zeng, T., Li, R., Mukkamala, R., Ye, J. and Ji, S., 2015. Deep convolutional neural networks for annotating gene expression patterns in the mouse brain. *BMC Bioinformatics*, 16, 147.

Zhao, X., Shi, P., Zheng, X. and Zhang, J., 2016. Intelligent tracking control for a class of uncertain high-order nonlinear systems. *IEEE Transactions on Neural Networks and Learning Systems*, 27 (9), 1976–1982.

Zhong, Z., Zheng, L., Kang, G., Li, S. and Yang, Y., 2017. Random erasing data augmentation. Available from https://arxiv.org/abs/1708.04896.

See Challenges as Opportunities for New Afresh Beginning....

5 Hybrid Fuzzy Classifiers

5.1 INTRODUCTION

Hybridization of classifiers begins with the fact that, in most cases of real world imagery such as remote sensing, no single classifier can provide the best result for all classes of ground features. For example, the traditional maximum likelihood classifiers (MLC) by virtue are based on Gaussian probability density modeling of the earth features. In many homogeneous objects, this is still a valid assumption. The benefits of these algorithms are fast to compute the required statistical parameters and compare with those in the library. On the other hand, artificial neural networks solve complex mixed classes that are common in high resolution imagery more efficiently than the MLC. However, the design is complex, and much trial and error would be required to freeze the right network design that yields good results for all the classes of interest. However, a combination of two classifiers is likely to benefit from the best of each when they complement each other, thereby yielding the best result. This hybridization can be at the pixel level, feature level, or decision level.

In this chapter, we describe fuzzy hybridization techniques wherein we use different properties of local features present in the image for improved classification applications. The properties of local features can be measured with a variety of functions. Some of the most common features are described as below:

1. Markov random field (MRF) is an approach widely used for characterizing contextual information. Contextual can be understood as correlation, and in this chapter spatial correction has been considered.
2. Entropy in information technology defines uncertainty, and this uncertainty concept has been applied to study the effect of entropy during classification.
3. Similarity/dissimilarity measures with many types of distance norms and can be applied in classifiers using distance function.
4. Spectral information divergence (SID) is a random probability distribution measure of classification that matches pixels in the image to the reference spectra, and for doing so, it utilizes the divergence function.
5. Spectral angle mapper (SAM) based spectral classifier uses angles to match unknown pixels to reference spectra.

5.2 HYBRID APPROACH

The purpose of the hybrid approach is to include different properties of images as input tools for the classification. The hybrid classification approach makes data consistent with two (or more) pieces of information of the image. The hybrid approach has been applied in different areas to bring two or more properties as input to generate more robust outputs. In many areas of remote sensing image classification, such a situation is generally faced when these important properties together can generate the better classified output. Thus, hybrid classifiers are generated using the different combination of various complementary heterogeneous methods. Classification is achieved by the joint effect of different methods after solving different types of tasks by each.

In this chapter, hybrid classifiers have been discussed in which contextual or entropy methods have been added with fuzzy classifiers. The fuzzy classifiers considered are FCM, PCM, and NC while contextual information has been used in the form of Markov random field (MRF). MRF has been initially developed to process the time series data. Here MRF has been applied to incorporate spatial contextual information. The spatial contextual information is equivalent to correlation between two neighbor pixels. Spatial correlation can be incorporated using local convolution approaches. In this chapter, spatial correlation has been explained to add through MRF as well as local convolution methods to generate hybrid classifiers. The entropy concept has also been included. Crisp *c*-means are fuzzified by various methods. The standard method of fuzzification given by Dunn and Bezdek in 1984 has the non-linearity $(\mu_{ki})^m$, and it also smoothes the solution from crisp into a differentiable one. Moreover, as "m" tends to 1, the fuzzy solution results toward the crisp solution. Therefore, it can be mentioned that the fuzzy based solution "regularizes" the crisp solution. These types of regularization are similar to the regularization faced while formulating problems having more than one solution. The function for regularization can be added in fuzzy based objective function as in Equation (5.1).

$$J'(U,V) = J_0(U,V) + \vartheta K(\mu), \vartheta > 0 \tag{5.1}$$

In Equation (5.1), $K(\mu)$ and ϑ are non-linear regularizing function and regularizing parameter respectively. The function represented by Equation (5.1) is known as the regularization by entropy method. Two types of the regularizing functions have been defined, an entropy function and a quadratic function, in Equations (5.2) and (5.3).

$$K(\mu) = \sum_{i=1}^{c} \sum_{k=1}^{N} \mu_{ki} \log \mu_{ki} \tag{5.2}$$

$$K(\mu) = \frac{1}{2} \sum_{i=1}^{c} \sum_{k=1}^{N} \mu_{ki}^2 \tag{5.3}$$

Due to convexity of these functions, these are capable to generate fuzzy membership matrix, with robust unique solutions. Miyamoto and Mukaidono (1997) first proposed this formulation using the maximum entropy method, which later has been reformulated by Li and Mukaidono (1999) using regularization function. The entropy based objective function has been developed using entropy function or quadratic function.

Before understanding fuzzy based objective functions, let us begin with the meaning of entropy in image processing. Entropy in the context of image shows uncertainty in the processed output from image. It is the amount of information content within the image. Digital images with low entropy values will, in many cases, have minimum contrast, or in other words, more pixels with similar DN values. A perfectly flat image with maximum homogeneity will have least entropy. Therefore, such low entropy images will have more redundancy with small DN range. On the contrary, high entropy images have high heterogeneity; therefore compression size is large in comparison to low entropy images. In the coming sections, soft fuzzy classifiers with entropy non-linear regularizing functions have been defined.

5.2.1 Entropy Based Hybrid Soft Classifiers

According to Shannon (1948, 1951), the measure of information or the information content has an intimate relationship with the entropy theory, as in statistical thermodynamics. Therefore, the information theory and thermodynamics must have some common points of interest. The increase in entropy has been regarded as the degradation of energy by Kelvin (Francis, 2000). In statistical thermodynamics, entropy is defined as a measure of the disorder of a system. However, in information theory, low entropy is a measure of the lack of information about the actual structure of the system (Feyerabend, 2015). It is thus perceived that fuzzy based information can become complete while adding entropy to the base classifier function. Due to the use of error function, the entropy based methods are similar to the statistical method. On the other hand, the standard soft classification methods such as FCM and PCM are different from the statistical method. Therefore, the standard methods are purely fuzzy, whereas entropy based methods are robust to both statistical and fuzzy models (Dunn, 1973; Bezdek, 1981). Thus, there is a motivation to hybridize the FCM and NC based classifiers with entropy. It is to be mentioned that the optimization of objective function with respect to membership values and cluster center as well as constraints are same for both cases, viz., with or without the entropy, whereas the difference is due to use of objective function. In the following sections, entropy based regularization function has been added with objective functions of FCM and NC fuzzy classifiers.

5.2.2 Fuzzy *c*-Means with Entropy (FCME)

Fuzzy *c*-means with entropy (FCME) is a hybridization approach of classification where the emphasis is to integrate the entropy based regularization method with FCM. According to various researchers, the standard FCM method becomes complete by adding entropy (Dunn, 1973; Bezdek et al., 1984; Miyamoto and Mukaidono, 1997; Li and Mukaidono, 1999). The primary motivation is to use both alternatives for the purpose of optimization considering membership matrix and cluster centers. Secondly, constraints are same for the both methods; the only difference is due to the objective functions. It is observed that the standard method of FCM initially proposed by Dunn (1973) and later adopted by Bezdek (1981) is fuzzy based. The entropy based FCM method follows the statistical models approach (Tihonov and Arsenin, 1997; Vapnik, 1995, 1998, 1999, 2000).

Further, from the experiments it has been noticed that output from standard FCM classifier gives better classification accuracy but less robust however, the output from entropy based FCM classifier, with optimum value of regularizing parameter, generates classified output with minimum uncertainty (Kumar and Dadhwal, 2010).

The objective function for FCME approach has been mentioned in Equation (5.4):

$$J_{\text{fcme}}(U,V) = \sum_{i=1}^{c}\sum_{k=1}^{N}(\mu_{ki})^m D(x_k, v_i) + \vartheta \sum_{i=1}^{c+1}\sum_{k=1}^{N} \mu_{ki}\log \mu_{ki} \tag{5.4}$$

where:

$U = N \times (c+1)$ matrix,
$V = (v_1 \ldots v_c)$ and
ϑ is the regularizing parameter.

From the objective function of FCME, the membership value can be calculated from Equation (5.5) and fuzzy mean from Equation (5.6):

$$\overline{\mu}_{ki} = \frac{\exp\left(-\frac{D(x_k, \overline{v}_i)}{\vartheta}\right)}{\sum_{j=1}^{c}\exp\left(-\frac{D(x_k, \overline{v}_j)}{\vartheta}\right)}, \quad 1 \le i \le c \tag{5.5}$$

$$\overline{v}_i = \frac{\sum_{k=1}^{N}\overline{\mu}_{ki} x_k}{\sum_{k=1}^{N}\overline{\mu}_{ki}}, \quad 1 \le i \le c \tag{5.6}$$

In Equation (5.4), the first term is the objective function of FCM classifier and second term is a non-linear regularizing entropy function. It is noticed that regularizing function is strictly a convex function and hence it is capable to generate the membership values.

5.2.3 Noise Clustering with Entropy (NCE) Classifier

Cluster analysis is used as an important tool for the satellite image data classification and analysis. Generally, it has been found that the c-means classifier, having c cluster centers, is still popular in remote sensing community. A collection of data points is gathered around a cluster mean and which generates a cluster, providing a base for a noise clustering algorithm. From entropy theory, given by Shannon (1948), entropy of systematic arranged data points is less while entropy of unsystematic arranged data points is higher. In a systematic arrangement dataset, maximum points are close to each other, while in unsystematic arrangement most of the data points are scattered randomly. Therefore, data point with minimum entropy is a good candidate for the information class (or cluster) center (Yao et al., 2000). In case of noisy data being present, noise has to be removed before calculating class center. However, in fuzzy clustering, entropy is directly evaluated with the least entropy data point as class center. Then, class centers with similar points within the threshold are removed (Yao et al., 2000; Chattopadhyay et al., 2011). Similarly, other class centers are selected from the remaining data points. Based upon the ideas of entropy with fuzzy clustering, Miyamoto et al. (2008) suggested to assign m (as defined in NC) as "1" and to add another term to consider entropy $K(u)$ along with a regularizing parameter (ϑ) to the objective function of NC to generate results of the noise clustering with entropy (NCE). It may be noted that, in the entropy based approach, the fuzzification is carried out with entropy and not by m. The objective function for NCE approach can be expressed by Equation (5.7):

$$J_{\text{nce}}(U,V)=\sum_{i=1}^{c}\sum_{k=1}^{N}(\mu_{ki})D(x_k,v_i)+\sum_{k=1}^{N}(\mu_{k,c+1})\delta+\vartheta\sum_{i=1}^{c+1}\sum_{k=1}^{N}\mu_{ki}\log\mu_{ki} \tag{5.7}$$

where:

$U=N\times(c+1)$ matrix,
$V=(v_1\ldots v_c)$ and
ϑ, the regularizing parameter.

By partial derivative of the objective function of the NCE, the membership values and fuzzy cluster center can be computed as Equations (5.8–5.10):

$$\bar{\mu}_{ki}=\frac{\exp\left(-\dfrac{D\left(x_k,\bar{v}_i\right)}{\vartheta}\right)}{\displaystyle\sum_{j=1}^{c}\exp\left(-\dfrac{D\left(x_k,\bar{v}_j\right)}{\vartheta}\right)+\exp\left(-\dfrac{\delta}{\vartheta}\right)},\quad 1\le i\le c \tag{5.8}$$

$$\bar{\mu}_{k,c+1} = \frac{\exp\left(-\frac{\delta}{\vartheta}\right)}{\sum_{j=1}^{c} \exp\left(-\frac{D\left(x_k, \bar{v}_j\right)}{\vartheta}\right) + \exp\left(-\frac{\delta}{\vartheta}\right)}, \quad 1 \le i \le c \tag{5.9}$$

$$\bar{v}_i = \frac{\sum_{k=1}^{N} \bar{\mu}_{ki} \cdot x_k}{\sum_{k=1}^{N} \bar{\mu}_{ki}}, \quad 1 \le i \le c \tag{5.10}$$

Equations (5.8) and (5.9) represent the membership value for information class and noise class, respectively. In Equation (5.7), the first and second terms are similar to the formulation of typical NC classifier. Further, the third term is a special regularization, while adding typical entropy as a linear regularizing function in the process of classification. The advantage of noise clustering is that untrained classes as noisy pixels are separated out quantitatively. Due to this capability of noise clustering, NCE can identify single class from a given image. All these algorithms use Euclidean distance, commonly. In the next section, large numbers of distances have been mentioned. These distances of different categories can be used with these classifiers.

5.3 SIMILARITY/DISSIMILARITY MEASURES IN FUZZY CLASSIFIERS

For the classification of any remote sensing image, training data points are always required. Sometimes the relationship within the training data points is also important. Similarity and dissimilarity measures are used to establish relation between the training data points for classification and clustering (Von Luxburg, 2004). Due to the wide range of data used in various technologies right from earth's surface to revolving satellites in space, it is very important to have information about coherence, redundancy, and degree of association of data that helps us in interpretation and meaningfulness of data. When we talk about the association of data, it can be interpreted as providing information about how similar and dissimilar data is to each other. The idea of similarity and dissimilarity is totally linked with each other and interdependent in terms as "degree of proximity between two points." Similarity functions are supposed to increase when the data points are more similar to each other; conversely, dissimilarity functions tend to increase when the data points are more dissimilar to each other. Therefore, we can say that similarity between two objects should grow as the dissimilarity between the objects decreases (Von Luxburg, 2004); that means they are complementary to each other. In the following sections, similarity and dissimilarity measures have been mentioned. These measures can be used as distance variables in fuzzy based classifiers mentioned in this book.

5.3.1 Similarity Measures

Let us consider two sets of orders of measurement $X = \{x_i: i = 1,2,3,4\ldots,n\}$ and $Y= \{y_i: i = 1,2,3,4\ldots,n\}$. The similarity between these two sets is a measure that brings the reliance between these X and Y to signify quantities from two objects or phenomenon. In remote sensing these sets can be assumed as two images, whereas x_i and y_i are digital numbers of corresponding pixels in the image. For a 2D image, these sequences can be according to the intensities in raster scan order. These similarity functions are called affinity functions. If a similarity measure generates a higher value as reliance between the intensities of corresponding pixels in the sequence increases, then it will be considered as a metric similarity measure. Metric similarity measures satisfy the following conditions (Theodoridis and Koutroumbas, 2006; Goshtasby, 2012):

1. *Limited Range*: $S(X, Y) \leq S_0$, for some arbitrarily large number S_0.
2. *Reflexivity*: $S(X, Y) = S_0$ if and only if $X = Y$.
3. *Symmetry*: $S(X, Y) = S(Y, X)$.
4. *Triangle Inequality*: $S(X, Y)\, S(Y, Z) \leq [Z(X, Y) + S(Y, Z)]\, S(X, Z)$.

where S_0 is the maximum similarity degree between all possible X and Y sequences. There are two other similarity measures (cosine and correlation) which can be used in distance based other classifiers like fuzzy based classifiers in single as well as in composite mode.

5.3.1.1 Cosine Similarity Measure

This calculates similarity between two vectors with an inner product space and generates cosine of the angle between them. Cosine values range between 1 and 0, and provide an orientation and not the magnitude. The cosine similarity of two vectors having the same orientation will be 1; two vectors orthogonal to each other will have a cosine similarity of 0. Generally, cosine similarity is used on the positive side with range between 0 and 1. There is no limitation of dimension, and therefore cosine similarity is mainly applied to the high-dimensional positive side. In an example of text data mining, each data is notionally given a different dimension and a document is represented by a vector where the value of each element represents the number of times that data appears in the document. From cosine similarity, it can be identified how similar two documents are to each other in terms of their subject matter. The technique is also used to find out similarity within clusters in the area of data mining. Cosine similarity is mathematically generated by Equation (5.11) (Ye, 2011):

$$D(X_j, V_i) = 1 - \frac{X_{j1}V_{i1} + X_{j2}V_{i2} + \ldots + X_{jb}V_{ib}}{\sqrt{Abs[X_{j1}]^2 + \ldots + Abs[X_{jb}]^2} \times \sqrt{Abs[V_{i1}]^2 + \ldots + Abs[V_{ib}]^2}} \tag{5.11}$$

5.3.1.2 Correlation Similarity Measure

This similarity measure between two vectors is calculated by the Pearson correlation coefficient. This coefficient was first developed by Karl Pearson in the 1880s. Since a correlation is a similarity function, it measures similarity rather than distance or dissimilarity. Its value range is between −1 and +1, where +1 represents perfect positive correlation, −1 as perfect negative correlation, and 0 as no correlation. It is considered a normalized form of covariance. Correlation similarity measure is given by Equation (5.12) (Zhang et al., 2008).

$$D(X_j, V_i) = 1 - \frac{\begin{bmatrix} \{X_{j1} + \frac{1}{b}(-X_{j1} - X_{j2} \ldots - X_{jb})\}\{V_{i1} + \frac{1}{b}(-V_{i1} - V_{i2} \ldots - V_{ib})\} + \ldots \\ + \{X_b + \frac{1}{b}(-X_{j1} - X_{j2} \ldots - X_{jb})\}\{V_b + \frac{1}{b}(-V_{i1} - V_{i2} \ldots - V_{ib})\} \end{bmatrix}}{\sqrt{Abs\left[X_{j1} + \frac{1}{b}(-X_{j1} - X_{j2} \ldots - X_{jb})\right]^2 + \ldots + Abs\left[X_b + \frac{1}{b}(-X_{j1} - X_{j2} \ldots - X_{jb})\right]^2} \times \sqrt{Abs\left[V_{i1} + \frac{1}{b}(-V_{i1} - V_{i2} \ldots - V_{ib})\right]^2 + \ldots + Abs\left[V_b + \frac{1}{b}(-V_{i1} - V_{i2} \ldots - V_{ib})\right]^2}} \tag{5.12}$$

where b denotes the number of bands in the image.

Pearson correlation between the two variables, let us say X and Y, is the covariance of these two variables divided by the product of their standard deviations. The case of perfectly positive correlation ($D = 1$) happens when X and Y coincide seamlessly; seamlessly negative correlation ($D = -1$) occurs when X and Y are completely out of phase, i.e., opposite to each other; and no correlation ($D = 0$) case means that variables X and Y are completely independent of each other.

5.3.2 Dissimilarity Measures

Let us again consider two sets of measurement $X = \{x_i: i = 1,2,3,4\ldots,n\}$ and $Y = \{y_i: i = 1,2,3,4\ldots,n\}$. The dissimilarity measure between these two sets represents a measure of independency between X and Y. The measure of dissimilarity here symbolizes the magnitude (or amounts) from two items or phenomena. If a dissimilarity measure D generates a highly less dependent value as a corresponding value between two sequences, then it will be considered as a metric dissimilarity measure. A metric dissimilarity measure satisfies the following conditions for all sequences of X and Y (Duda et al., 2001; Theodoridis and Koutroumbas, 2006; Goshtasby, 2012):

1. *Non-negativity*: $D(X, Y) \geq 0$.
2. *Reflexivity*: $D(X, Y) = 0$ if and only if $X = Y$.
3. *Symmetry*: $D(X, Y) = D(Y, X)$.
4. *Triangle Inequality*: $D(X, Y) + D(Y, Z) \geq D(X, Z)$.

The dissimilarity measures such as *Euclidean, Manhattan, Canberra, chessboard, Bray Curtis, mean absolute difference, median absolute difference*, and *normalized squared Euclidean* have been discussed in the following section.

5.3.2.1 Euclidean Distance

In spectral similarity, Euclidean Distance (ED) calculates similarity between two spectral signatures by calculating square root of the squared difference between the two data points. If s and s' are two spectral signature vectors of two pixels where "n" is the set of spectral band channels, then Euclidean measure can be given as stated in Equation (5.13):

$$ED(s,s') = |s - s'| = \sqrt{\sum_{j=1}^{n} \left(s_j - s'_j\right)^2} \tag{5.13}$$

However, in a metric space the Euclidean distance is basically the normal distance between two objects (generally two pixel values). This measure was first introduced by Bezdek (1984). As the measure has to operate on images, it was introduced in a form of identity matrix. The ED measure is given as stated in Equation (5.14): Other than simple normal distance, two other norms of distance, viz. diagonal and Mahalanobis, can be considered depending upon matrix "A" type to be diagonal variance–covariance or variance–covariance Equation (5.14).

$$D(X_j, V_i) = \left\| X_j - V_i \right\|_A^2 = (X_j - V_i)^T A (X_j - V_i) \tag{5.14}$$

The matrix A can take different norms as discussed in Chapter 3.

5.3.2.2 Manhattan Distance

This is the sum of absolute intensity differences and is one of the oldest dissimilarity measures used to compare images. In other words, the Manhattan distance between two vectors is the sum of the differences of their corresponding components. This function computes the grid-like path distance from one data point to another. This norm is also called the *city block distance* or *taxicab distance*; it is called so because of its use in the calculation of shortest distance path which a car could take between two intersections in the borough to have length equal to the intersection's distance in the Manhattan city block. Manhattan distance is a slight variation of Euclidean distance and it differs in a way of calculating distance between two data points. If vector pixel value $X_j = (X_{j1}, X_{j2}, X_{j3},\ldots X_{jn})$ and mean value is $V_i = (V_{i1}, V_{i2}, V_{i3},\ldots,V_{in})$, then Manhattan distance will be given as in Equation (5.15) (Hasnat et al., 2013)

$$D\left(X_j, V_i\right) = Abs\left(X_{j1} - V_{i1}\right) + Abs\left(X_{j2} - V_{i2}\right) + \ldots + Abs\left(X_{jn} - V_{in}\right) \tag{5.15}$$

In generalized form it is given by Equation (5.16):

$$D\left(X_j, V_i\right) = \left[\sum_{i=1}^{n} Abs\left(X_j - V_i\right)\right] \tag{5.16}$$

5.3.2.3 Chessboard

A metric chessboard distance is a vector space wherein the distance between two vectors is the greatest of their distances along any coordinate dimension (Cantrell, 2000; Abello et al., 2013). It is also called Chebyshev distance as it was named after mathematician Pafnuty Chebyshev. It is popularly known as chessboard distance, inspired by the game of chess where the minimum moves are required by a king from one square to another square on a chessboard. This distance is equal to Chebyshev distance between the centers of the squares, if the squares have side length one, as represented in 2D spatial coordinates with axes aligned to the edges of the board (Van Der Heijden et al., 2005). The chessboard distance may be suitable if the difference between vectors is reflected more by differences in individual dimensions rather than all the dimensions considered together. The chessboard distance is given by Equation (5.17):

$$D\left(X_j, V_i\right) = \text{Max}\left[Abs\left(X_j - V_i\right)\right] \tag{5.17}$$

5.3.2.4 Bray Curtis

Bray Curtis is a statistical dissimilarity and named after J. Roger Bray and John T. Curtis. Apart from image processing, it is also used in biology and ecology. It is used to quantify the compositional dissimilarity between two different sites, based on counts at each site. Bray Curtis is a normalization method that is commonly used in botany, ecology, and environmental science fields. It visualizes the space as a grid just like city block distance. It is directly related to the Sorenson similarity index. One important property: if all the coordinates are positive, then the Bray Curtis value is between 0 and 1. For exactly similar coordinates, its value is 0. If either vectors overlap or have zero coordinates, then Bray Curtis dissimilarity cannot be defined (Bloom, 1981). The Bray Curtis dissimilarity does not satisfy the triangle inequality and hence it is not exactly a distance. Bray Curtis dissimilarity is given by Equation (5.18):

$$D\left(X_j, V_i\right) = \Sigma Abs\left[X_j - V_i\right] \Sigma Abs\left[X_j - V_i\right]; \left(\text{Bray Curtis Index}\right) \tag{5.18}$$

5.3.2.5 Canberra

Canberra distance is an absolute numerical measure of distance between pairs of vectors in a vector space (Lance and Williams, 1966, 1967). It examines the sum of series of fraction differences between coordinates of a pair of objects. This distance has application in comparing ranked lists (Jurman et al., 2009) and for intrusion detection in computer security (Emran and Ye, 2002). Canberra distance is equivalent to weighted version of Manhattan distance (Jurman et al., 2009) and is given by Equation (5.19):

$$D\left(X_j, V_i\right) = \Sigma Abs\left[X_j - V_i\right] / \left[\Sigma Abs\left[X_j\right] + Abs\left[V_i\right]\right] \tag{5.19}$$

5.3.2.6 Mean Absolute Difference

This is a measure of statistical diffusion, equivalent to average absolute difference of two independent random vectors drawn for a probabilistic distribution function. Mathematically, it is the mean of absolute non-conformities from the central point of the data sets. It is given by Equation (5.20) (Vassiliadis et al., 1998):

$$D(X_j, V_i) = \frac{1}{b}\left[Abs(X_{j1} - V_{i1}) + Abs(X_{j2} - V_{i2}) + \ldots + Abs(X_{jb} - V_{ib})\right] \quad (5.20)$$

5.3.2.7 Median Absolute Difference

For an impulse noise or due to the presence of salt and pepper noise, Manhattan norm produces an exaggerated distance measure. Let us consider the fixed size images having n pixels each, then according to Manhattan norm the sum of absolute intensity difference will be equivalent to the average absolute intensity difference for corresponding pixels. Instead of using average (mean) absolute difference, to reduce the effect of impulse noise, median of absolute differences may be used to measure the dissimilarity between two vectors. The median absolute difference can be presented by Equation (5.21) (Scollar et al., 1984):

$$D\left(X_j, V_i\right) = \text{Median}\left(X_j - V_i\right) \quad (5.21)$$

5.3.2.8 Normalized Squared Euclidean

This norm standardizes the measure with respect to image minimum and maximum differences. It is slightly slower in comparison to correlation coefficient similarity measure, because it requires standardization of intensity before calculating the sum of squared differences between them. For the calculation of correlation coefficient, scale normalization is performed once after calculating the inner product of the normalized intensities. The normalized squared Euclidean distance is computed by Equation (5.22):

$$D(X_j, V_i) = \frac{\begin{array}{l} Abs\left\{X_{j1} + \frac{1}{b}(-X_{j1} - X_{j2} \ldots - X_{jb}) - \text{V}_{i1} + \frac{1}{b}(V_{i1} + V_{i2} \ldots + V_{ib})\right\}^2 + \ldots \\ + Abs\left\{X_{jb} + \frac{1}{b}(-X_{j1} - X_{j2} \ldots - X_{jb}) - \text{V}_{ib} + \frac{1}{b}(V_{i1} + V_{i2} \ldots + V_{ib})\right\}^2 \end{array}}{2\left[\begin{array}{l} Abs\left\{X_{j1} + \frac{1}{b}(-X_{j1} - X_{j2} \ldots - X_{jb})\right\}^2 + \ldots \\ + Abs\left\{X_{jb} + \frac{1}{b}(-X_{j1} - X_{j2} \ldots - X_{jb})\right\}^2 + \\ Abs\left\{\text{V}_{i1} + \frac{1}{b}(-V_{i1} - V_{i2} \ldots - V_{ib})\right\}^2 + \ldots + Abs\left\{\text{V}_{ib} + \frac{1}{b}(-V_{i1} - V_{i2} \ldots - V_{ib})\right\}^2 \end{array}\right]} \quad (5.22)$$

5.3.2.9 Composite Measure: Combining Similarity and Dissimilarity Measures

Composite measure is a term for the weighted combination of similarity and dissimilarity measures. It can be developed with the help of two measures giving certain weight to each similarity and dissimilarity component. An example of composite measure is given by Equation (5.23)

$$D_c = \lambda D_a + (1-\lambda) D_b \tag{5.23}$$

where D_c is a composite measure, D_a and D_b can be any similarity and dissimilarity measure, and λ is a weighting component, $0 \le \lambda \le 1$.

5.4 SPECTRAL CHARACTERIZATION MEASURES

Three principal spectral characterization measures can also be included for use in a fuzzy based classifier. Spectral characterization measures have been previously worked on hyperspectral data for various applications, however their performance on a multi-spectral image with lower spectral resolution needs to be explored. It is obvious that spectral characterization measures work well on the data with high spectral resolution. Additionally, a concept of measure hybridization has been explored which has proven to be effective for many applications like geology and crop identification. The spectral characterization measures covered in this section are singular measures like spectral information divergence (SID), spectral angle mapper (SAM), and spectral correlation angle (SCA)

5.4.1 Spectral Information Divergence (SID)

Spectral information divergence (SID) is a stochastic measure for classification that matches pixels in the image to the reference spectra and by doing so it utilizes divergence function. It calculates the distance between the probability distributions produced by the spectral signatures of two pixels (Chang, 2000). For a pixel vector, it models the probability distribution of its spectrum. It was framed on the basis of spectral information measure (SIM), which uses spectral signature histogram and considers each pixel to be a random variable. It is defined for two pixels s_i and s_j as stated in Equations (5.24–5.26):

$$\text{SID}(s_i, s_j) = D(s_i \| s_j) + D(s_j \| s_i) \tag{5.24}$$

$$D(s_i \| s_j) = \sum_{l=1}^{L} p_l D_l(x \| y) = \sum_{l=1}^{L} p_l(I_l(x)) - (I_l(y)) \sum_{l=1}^{L} p_l \log \frac{p_l}{q_l} \tag{5.25}$$

and

$$D(s_j \,||\, s_i) = \sum_{l=1}^{L} q_1 \log \frac{q_1}{p_1} \tag{5.26}$$

where p_l and q_l are probability vectors of the spectral signatures of vectors s_i and s_j. $I_1(x)$ and $I_1(y)$ stand for entropy of the information.

5.4.2 Spectral Angle Mapper (SAM)

Mostly used for hyperspectral remote sensing, spectral angle mapper (SAM) is a spectral similarity measure that calculates the angle between the spectral signatures of two samples. This measure calculates the angle between the two vectors using the same formula that is used in calculating the angle between two vectors (Van der Meer, 2006), i.e., spectra are supposed to behave like vectors. For signatures s_i and s_j, the SAM is defined as stated in Equation (5.27) and angle from Equation (5.28):

$$\text{SAM}(s_i, s_j) = \cos^{-1}\left(\theta_{s_i, s_j}\right) \tag{5.27}$$

where:

$$\theta_{s_i, s_j} = \left\{ \frac{\sum_{i,j=1}^{L} s_i s_j}{\sqrt{\sum_{i=1}^{L} s_i^2} \sqrt{\sum_{j=1}^{L} s_j^2}} \right\} \tag{5.28}$$

The two SAM based measures have been framed using the value of θ_{s_i, s_j} as stated in Equation (5.29):

$$\text{SAM} - \tan = \tan \theta_{s_i, s_j} \quad \& \quad \text{SAM} - \sin = \sin \theta_{s_i, s_j} \tag{5.29}$$

Unlike cosine, which projects one spectra along the other and reduces the discrimination of spectra, tan and sin increase the power of discrimination of the SAM measure.

5.4.3 Spectral Correlation Angle (SCA)

Spectral correlation angle (SCA) is a correlation function based on a deterministic measure. This measure computes a coefficient of correlation between two spectral signatures. Suppose we have two spectral signatures $s_i = (s_{il}, \ldots s_{it})^T$ and $s_j = (s_{jl}, \ldots s_{jt})^T$. Spectral correlation angle can be determined on basis of the Pearsonian correlation coefficient (Naresh Kumar et al., 2011). The Pearsonian coefficient is expressed as stated in Equation (5.30):

$$r_{s_i,s_j} = \frac{n\sum_{i,j=1}^{n} s_i s_j - \sum_{i=1}^{n} s_i \sum_{j=1}^{n} s_j}{\sqrt{\left[n\sum_{i=1}^{n}(s_i)^2 - \left(\sum_{i=1}^{n} s_i\right)^2\right]\left[n\sum_{j=1}^{n}(s_j)^2 - \left(\sum_{j=1}^{n} s_j\right)^2\right]}} \tag{5.30}$$

where r_{s_i,s_j} is the Pearsonian correlation coefficient and "n" is the number of spectral bands of the image.

The Pearsonian coefficient can take any value from −1 to +1 and portrays a linear interdependence between two spectra (s_i & s_j). It is a dimensional index that is converted to angular form using the formula as stated in Equation (5.31):

$$\text{SCA}\left(s_i, s_j\right) = \cos^{-1}\left(\frac{r_{s_i,s_j}+1}{2}\right)\ (\text{in radians}) \tag{5.31}$$

The spectral correlation angle is symmetric and invariant quantity to multiply with positive scalars and value of SCA ϵ [0, 1.570796].

5.5 HYBRIDIZATION OF SPECTRAL MEASURES

5.5.1 SID-SAM Hybridization

These hybrid measures were first proposed in 2004 for spectral characterization on hyperspectral data. These can also be deployed on multi-spectral data for the purpose of image soft classification. The SID-SAM hybrid measure has the property of making two distinct spectra more different and two similar spectra even more similar. The SID-SAM hybrid measure is combined with trigonometric functions tan and sin as they enhance the degree of discrimination among the spectra. SID-SAM between two spectral signatures s_i and s_j is defined as stated in Equations (5.32) and (5.33):

$$\text{SID-SAM} - \tan = \text{SID}(s_i, s_j) \times \tan\left(\text{SAM}(s_i, s_j)\right) \tag{5.32}$$

$$\text{SID-SAM} - \sin = \text{SID}(s_i, s_j) \times \sin\left(\text{SAM}(s_i, s_j)\right) \tag{5.33}$$

5.5.2 SID-SCA Hybridization

The SID-SCA hybrid measure is similar to SID-SAM but a few studies like Naresh Kumar et al. (2011) found this measure work better than SID-SAM with enhanced capabilities of spectra discrimination. The SCA has a capacity to point out false positive results in measurement of spectral properties and hence effectively works on hyperspectral images. These hybrid measures can be applied on a multi-spectral image and be used to deduce if it displays similar enhanced discriminatory powers on a multi-spectral image. The SID-SCA between two spectra can be stated as Equations (5.34) and (5.35):

$$\text{SID-SCA} - \tan = \text{SID}(s_i, s_j) \times \tan\left(\text{SCA}(s_i, s_j)\right) \tag{5.34}$$

$$\text{SID-SCA} - \sin = \text{SID}(s_i, s_j) \times \sin\left(\text{SCA}(s_i, s_j)\right) \tag{5.35}$$

5.6 KERNELS CONCEPT IN FUZZY CLASSIFIERS

The kernel concept is generally used in machine learning for handling non-linearity, especially in support vector machine (SVM) classifiers. In the kernel concept, a linearly separating optimal hyperplane is fitted between the different class feature vectors in higher dimensional feature space (Camps-Valls and Bruzzone, 2009). Samples are separated in a manner that identical classes will be grouped as one side of the hyperplane. According to Boser et al. (1992), it is better to maximize the margin between two class boundaries for optimizing the cost functions such as mean square error. In a data set, if classes are non-linear, then the data has to be transformed to a higher dimensional space where they are considered to be linearly separable with the help of kernel (Figure 5.1).

Let us illustrate the kernel mapping by considering some training data of two given sets $S \times T$ such that (Camps-Valls and Bruzzone, 2009) Equation (5.36):

$$(s_1, t_1)(s_2, t_2)\ldots(s_n, t_n) \in S \times T \tag{5.36}$$

Here s_i represents input data from a set S, and t_i represents the target elements from set T. Original input sample set S is transformed into higher dimensional feature space, Equations (5.37) and (5.38).

$$\phi : S \rightarrow F, s \leftrightarrow \phi(s) \tag{5.37}$$

Let us consider samples l and l_i in the input space, then

$$K(s, s_i) = \langle \phi(s), \phi(s_i) \rangle_F \tag{5.38}$$

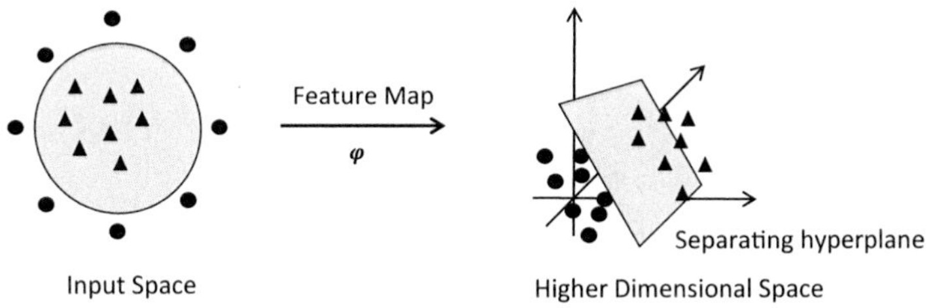

FIGURE 5.1 Data in higher dimensional space through kernel.

Here the function K is a kernel function while $\langle \phi(s), \phi(s_i) \rangle$ is the inner product between s and s_i. Mapping function ϕ is referred to as the feature map and F is the feature space (Camps-Valls and Bruzzone, 2009). Use of such kernel function reduces the computational complexity in the original input space by a considerable amount. Mercer's condition for kernels states that [Equation (5.39)]:

$$K\left(x, x_i\right) \geq 0 \tag{5.39}$$

A kernel that satisfies the Mercers condition is called an *eligible kernel* (Kumar, 2007). Mercer condition for a function $K\left(x, x_i\right)$ is analogous to the non-negative definiteness for a matrix. Linearly separable classes are the simplest case to train a support vector machine. Boser et al. (1992) proposed that data is transformed into a higher dimensional Euclidean space (feature space) F via a non-linear vector function; however, computationally it is time consuming. To cope up with this problem, Cortes and Vapnik (1995) introduced the concept of the kernel function K in the design of non-linear SVMs.

In SVM algorithm, the kernel functions are constructed while mapping the data into a higher dimensional space in absence of any possibility of linear separation in the original one. This approach consists of transforming the data in a higher dimensional space to make it linearly separable. On higher dimensional data, any classifier can be applied to divide data into different classes as data has been transformed to higher dimensions. It can also be generalized to compute non-linear decision surfaces. Large numbers of kernels exist, and it is difficult to explain their individual characteristics. Some common kernels used for handing various types of data are categorized as local kernels, global kernels, and spectral kernels.

5.6.1 Local Kernels

In these kernels, only the close neighbor or the proximity local data has an influence on the kernel values. Basically, all distance kernels are local kernels. Some listed local kernels are as follows in Equations (5.40–5.43):

Gaussian kernel (Refaat and Farag, 2004):

$$K\left(x, x_i\right) = \exp\left(-0.5\left(x - x_i\right) A^{-1} \left(x - x_i\right)^T\right) \tag{5.40}$$

where "A" is a weight matrix known that can have different norms, Euclidean, diagonal, and Mahalanobis, as discussed previously.

Radial basis:

$$K(x, x_i) = \exp(- \| x - x_i \|^2) \tag{5.41}$$

Kernel with moderate decreasing (KMOD):

$$K(x, x_i) = \exp\left(\frac{1}{1 + \| x - x_i \|^2}\right)^{-1} \tag{5.42}$$

Inverse multi-quadric:

$$K(x, x_i) = \frac{1}{\sqrt{(\| x - x_i \|^2 + 1)}} \tag{5.43}$$

5.6.2 Global Kernels

When the samples are far away from each other and still influencing the kernel value, such kernels are known as global kernels. Those kernels that follow dot product are known as global kernels. Some of the global kernels are given in Equations (5.44)–(5.46):

Linear:

$$K(x, x_i) = x \cdot x_i \tag{5.44}$$

Polynomial:

$$K(x, x_i) = (x \cdot x_i + 1)^p \tag{5.45}$$

Sigmoid:

$$K(x, x_i) = \tan h(x \cdot x_i + 1) \tag{5.46}$$

5.6.3 Spectral Kernels

Local kernels are based on a quadratic distance between two samples. For hyperspectral nature of the data, a new criterion that takes care of the spectral signature into consideration is of interest. In order to measure the spectral difference between x and x_i spectral angle $\alpha(x, x_i)$ can be used. This kernel is robust to the differences of the overall energy, such as illumination and shadow (Mercier and Lennon, 2003), Equations (5.47) and (5.48).

Spectral angle (SA):

$$\alpha(x, x_i) = \arccos\left(\frac{x \cdot x_i}{\| x \| \| x_i \|}\right) \tag{5.47}$$

Hyper tangent kernel:

$$K(x, x_i) = 1 - \tanh\left(-\frac{\|x - x_i\|^2}{\sigma^2}\right) \tag{5.48}$$

5.6.4 Hybrid Kernel Approach

The fusion of multi-spectral image with panchromatic image sharpens the resultant multi-spectral. In a similar manner, in case of kernel function, a mixture of kernels can be used to mix the dot product or Euclidean distance with the spectral angle (Mercier and Lennon, 2003; Kumar, 2007). The spatial and spectral characteristics of Mercer's single kernel can be combined to a new family known as composite kernels. This makes the classification more flexible by taking advantage of both the spatial and spectral properties and also increasing the computational efficiency (Camps-valls et al., 2006). The methods for combining two different kernels include cross-information kernel, direct summation kernel, stacked approach, and weighted summation kernel. The weighted summation kernel method is generally adopted by various researchers. In this method the composite kernel can be expressed as in (Kumar, 2007) Equation (5.49):

$$K(x, x_i) = \lambda k_a(x, x_i) + (1 - \lambda) k_b(x, x_i) \tag{5.49}$$

where $k_a(x, x_i)$ and $k_b(x, x_i)$ are two local (or global) and spectral kernels respectively and λ represents the weight of the kernel with values between 0 and 1 ($0 < \lambda < 1$). Optimization of λ is also required with the degree of fuzziness for the composite kernels. If $k_a(x, x_i)$ and $k_b(x, x_i)$ both satisfy the Mercer's condition for eligible kernels, then the linear combination is also an eligible kernel. The best single kernel among the local as well as global category can be combined with the spectral kernel. $k_a(x, x_i)$ kernel can be taken to be any local or global kernel while $k_b(x, x_i)$ kernel has been taken as a spectral kernel.

These nine kernels in individual as well as in hybrid mode can be applied with fuzzy based classification algorithms as discussed in Chapter 3 and their hybrid versions in Chapter 5. While applying kernel in fuzzy classifiers, these algorithms can be mentioned as KFCM, KPCM, KNC, KIPCM, and KMPCM.

5.7 THEORY BEHIND MARKOV RANDOM FIELD (MRF)

Bayesian's network and the Markov random field (MRF) both are graph formalism methods, while modeling the joint probability distribution. The main difference between them is that the Bayesian's networks are directed while the MRF are undirected graphical models. MRF is an n-dimensional random process defined on a discrete lattice. Statistical modeling of spatial random field can easily be done using the MRF. A Markov network is based on Markovian property, generally referred to as the memory-less property of a stochastic process. The probability distribution function of future states of the process depends upon the present state, not on the sequences of preceding stages. Thus, there is a focus on a single random variable at a particular time and hence simulation of MRF becomes simple.

MRF is an undirected graphical model which explicitly exhibits the conditional independence relationship between nodes. MRF networks are similar to Gibbs random field in the sense of undirected path. The basic difference is in the form of probability distribution function. The Gibbs fields have implicit probability distribution function while MRF only specifies the conditional independence. Gibbs random field can be represented as Equation (5.50):

$$P \simeq \exp\left[\frac{-U(r)}{T}\right] \tag{5.50}$$

where P is the probability of r, $U(r)$ the energy function, and T the temperature.

5.7.1 MAP-MRF Framework

According to the Bayes theory, when both prior distribution and neighborhood function of a pattern are known, then Bayes labeling can easily be done. The maximum a posterior (MAP) proposed by Geman and Geman (1984) is a special case in the Bayes framework solution. The MAP-MRF framework searches for the most probable configuration for the class in the image. It is done by maximizing the posterior probability $P(x/d)$, Equations (5.51)–(5.54).

$$x = \arg\max_x P(x/d) \tag{5.51}$$

According to the Bayesian:

$$P(x/d) = \frac{P(d/x)}{P(d)} \tag{5.52}$$

The MAP estimate can be given as:

$$\hat{x} = \arg\min_x U(x/d) \tag{5.53}$$

or

$$\hat{x} = \arg\min_x \left[U(d/x) + U(x)\right] \tag{5.54}$$

where $\hat{x}$ is the optimal variable value in case of classification it is class label, $U(d/x)$ is the conditional energy, and $U(x)$ is the prior energy.

5.7.2 Contextual Information Using MRF

"He who loves practice without theory is like the sailor who boards ship without a rudder and compass and never knows where he may cast", said Leonardo Da Vinci. In addition to the DN value of pixels, any other information such as contextual information helps while interpreting the digital image. For example, in image labeling, if a pixel is in isolation this can lead to missing or incomplete information. The additional contextual information will be beneficial for the image interpretation and for improving the classification accuracy, while reducing noise. Contextual information can be acquired from any domain, viz., spatial, spectral, and temporal. MRF provides a logical and mathematically consistent way of establishing the contextual relationship among pixels of the image. Spatial contextual information has been used widely in image classification, image denoising, and image segmentation.

Contextual information is similar to correlation or homogeneity. Contextual information is initially generated in time domain using time series data. However, contextual information has potential to be used for spatial domain. Conventional fuzzy based classifiers are using spectral property of an image. Thus, standard fuzzy-set based classifiers algorithms do not incorporate spatial contextual information of the pixels. The spectral information of the pixels is not sufficient enough to handle noise, uncertainty, and vagueness in a class. Therefore, to incorporate spatial contextual information, Markov random field (MRF) is widely used (Solberg et al., 1996; Li, 2009). According to Li (2009), MRF theory is able to model the contextual based entities such as pixels or correlated features in a convenient and consistent way. Contextual information in spatial domain implies the correlation of class labels for neighboring pixels (Solberg et al., 1996). The geographical phenomenon is in the context to each other. As an example, a particular vegetation pixel will have a high probability to be the same vegetation pixel as its neighbors. Therefore, isolated or pepper and salt type of pixels exist rarely. Thus, applying contextual information removes isolated or so called pepper and salt pixels (Tso and Mather, 2009).

To incorporate contextual information, selection of MRF model is an important step. The MRF models are also known as MRF priors and regularizers. Examples of MRF models are standard regularization model, weak string and membrane model, line process model, and discontinuity adaptive (DA). The standard regularization model (i.e. smoothness prior) and DA models (i.e. edge preserving priors) are widely used to study smoothing as well as edge preserving effects in images. To model smoothness in an image, MRF uses smoothness prior models for calculating prior energy using prior probabilities (Tso and Mather, 2009). It applies smooth contextual concept, which assumes uniformity everywhere in the image.

MRF gives a helpful and conceptually well-established model to incorporate contextual information in the classification process (Melgani and Serpico, 2003). According to Solberg et al. (1996), MRF based contextual techniques can be utilized for fusion of multi-source information with enhanced classification accuracy. For dealing with a series of temporal images, the standard MRF based approach comprises of embracing a "cascade" scheme. According to Melgani and Serpico (2003), the MRF model, on one hand, exploits the spatial class relationship between neighboring pixels in an image and on the other hand, provides the temporal class relationship between different images of the same scene.

5.7.3 Contextual Fuzzy Classifier

Contextual fuzzy classifier can be developed using the MRF model with any fuzzy based classifier.

While including contextual information in fuzzy *c*-means classifiers, there are some issues mentioned as follows:

a. The summation of membership values for a pixel for all the classes should be up to 1.
b. The membership values for a single pixel for all the classes need to be updated at the same time. This makes the process more complex in comparison with hard contextual classifier where at a time for a given pixel only one class label is updated.
c. There can be an infinite number of membership values possible in form of real numbers between 0 and 1. Sampling from such space is computationally hard.

5.7.4 Smoothness Prior

To add contextual information, one of the methods is smoothness prior. To optimize global posterior energy function, simulated annealing (SA) can be applied. Simulated annealing (SA), a stochastic relaxation algorithm, can be used for optimization of global energy function. The concept to SA was initially used in thermodynamics and material science to remove stress from a system with a large number of particles. In simulated annealing, a material or any metal is heated to a desired temperature and then cooled at some rate to have enough time to respond. This concept is similar to adding some noise in a system and shakes the search process away from the local minima to get a global minimum. Initially Metropolis et al. (1953) have proposed the SA technique for sampling from a random field. It was reported that their technique performed better in comparison to Gibbs sampler. The mathematical form of the Metropolis algorithm can be given by, Equation (5.55):

$$U\left(\mu / d\right) = (1-\lambda)\sum_{i=1}^{c}\sum_{k=1}^{N}\mu_{ki}^{m}\sqrt{\sum_{l=1}^{L}\left(d_{kl}-v_{il}\right)^{2}} + \lambda\sum_{k=1}^{N}\left(1-\sum_{i=1}^{c}\sum_{j\in N_k}\beta\sqrt{\mu_{ki}\mu_{ji}}\right) \tag{5.55}$$

where:

$U\left(\mu/d\right)$ is the posterior energy of membership value μ, given observed image d.
λ is the smoothness parameter or weight factor for context and spectral information.
μ_{ki} is a class membership values of a pixel k to class i.
d_{kl} is the original DN value for a pixel k in band l.
v_{il} is the class mean value for class i for band l.

m is the fuzzy weight value.
β is the weight for neighbors.
L is the total number of bands in a multi-spectral image.
N_k is the neighborhood window around pixel k.
N is the total number of pixels in the image.

This algorithm generates the new membership image configuration μ'_{ki} through conditional random function and calculates the global posterior energy function associated with it. Further, lower global posterior energy is checked in comparison to previous energy. If this is true, then the new configuration accepts with probability 1, otherwise it is less than 1. But if a new configuration generates higher energy in comparison to the previous iteration, then it is accepted with a probability that decreases with the increment in the difference in energy states at previous and new configuration. In Equation (5.54), by taking $\lambda = 0$ results in a solution which is like FCM algorithm. This happens due to the fact that smoothness prior is ignored in energy expression. But practically, it has been observed that if $\lambda = 0$ the MRF-FCM solution is not equivalent to FCM solution. For this, Gibbs sampler can be used to resolve this deficiency in the developed MRF-FCM algorithm. The algorithm for the Gibbs sampler is given by Equation (5.56):

$$U(\mu/d) = (1-\lambda)\sum_{i=1}^{c}\sum_{k=1}^{N}\left[\mu_{ki} - \mu_{ki}^{\text{fcm}}(d_{kl})\right]^2 + \lambda\sum_{i=1}^{c}\sum_{k=1}^{N}\sum_{j\in N_k}\beta\left(\mu_{ki} - \mu_{ji}\right)^2 \quad (5.56)$$

where μ_{ki}^{fcm} is the solution of supervised FCM algorithm.

In Gibbs sampler, the pixel value is updated using Gibbs random field mentioned in Equation (5.50): $P \simeq \exp\left[\frac{-U(r)}{T}\right]$.

The algorithm starts with a high temperature, and once it converges, the temperature is decreased with a carefully defined cooling schedule until T becomes 0 similar to simulated annealing. Based on the local conditional distribution, Gibbs sampler generates the new membership value μ'_{ki} for each pixel k and class i. It also includes the brightness temperature T. The Gibbs sampler concept is similar to the Metropolis concept except the computation of new membership value μ'_{ki}.

5.7.5 Discontinuity Adaptive (DA) Priors

In a digital image, smoothness means that there is no abrupt change in the physical properties of the system. Thus, DN values remain nearly identical and do not change frequently in a particular part of image and remain coherent with each other. However, this remains an assumption, and in a practical scenario there is piecewise discontinuity for the DN values in an original satellite digital image.

MRF assumes the smoothness prior model; however, its improper imposition may lead to over-smoothening and hence undesirable results. This generally occurs when there is any discontinuity especially at class boundaries or edge. Therefore, it becomes important to select MRF models which can take into account the possible discontinuity. Discontinuity adaptive smoothness priors and regularizers are used to encode the prior energy. In MRF, the analytical regularizers are used; these include the prior constraints and penalize the irregularities. MAP-MRF framework is used to maximize the posterior probability and hence minimizes the corresponding posterior energy. A general form of these regularities is given by, Equation (5.57):

$$U(x) = \sum_{n=1}^{N} U_n(x) = \sum_{n=1}^{N} \lambda_n \int_a^b g\left[\left\{x^n(y)\right\}\right] dy \tag{5.57}$$

where $U(x)$ prior energy is called here n^{th} order regularizer and $g\left[\left\{x^n(y)\right\}\right]$ is the penalty function and is identical to clique potential.

5.7.5.1 Standard Regularization

These regularizers use the quadratic function in the form of $g_q(\alpha) = \alpha^2$. This implies that a higher irregularity in $x^{n-1}(y)$ at a site x will lead to higher value of $x^n(y)$ and hence $g\left[\{x^n(y)\}\right]$. This results in an increased value of energy $U(x)$. Such type of regularization can be used with FCM based classifier. Limitations of the standard regularization are as follows:

1. There is a constant interaction between neighborhoods.
2. The strength of smoothening is proportional to the derivative of magnitude of x^n.
3. Many times it may lead to over-smoothening at discontinuities.

5.7.5.2 DA MRF Model

Regularizers generally differ by method of interactions among their neighborhood and strength. To avoid over-smoothening, at discontinuities the interactions must be managed accordingly. This is a need catered by DA models. There are four choices of potential function $\left(\text{say } g(\alpha)\right)$ available, known as adaptive potential function (APF). The derivative of APF can be expressed as Equation (5.58):

$$g'(\alpha) = 2\alpha h_\gamma(\alpha) \tag{5.58}$$

where h_γ influences the interaction between the neighborhood sites and known as influence function. h_γ is termed as adaptive interaction function (AIF).

The strength with which a regularizer will perform smoothening is given by Equation (5.59):

$$|g'(x')| = |2x'h(x')| \tag{5.59}$$

where α is given as $\alpha = x'(y)$.

All aforementioned equations and DA models are derived from the Euler's equation, Equation (5.60):

$$u_x\left(x,x'\right)-\frac{d}{dx}u_{x'}\left(x,x'\right)=0 \tag{5.60}$$

5.7.5.3 How DA Priors Work

The necessary condition for a regularizer to be "discontinuity adaptive" is as in Equation (5.61):

$$\lim_{\alpha\to\infty}\left|g'\left(\alpha\right)\right|=\lim_{\alpha\to\infty}\left|2\alpha h\left(\alpha\right)\right|=C \tag{5.61}$$

where C is a constant and $C\in\left[0,\infty\right]$. If $C=0$, smoothening strength will be zero; for $C>0$ there will bound smoothening. The four possible choices for the DA models are as in Equations (5.62–5.65):

$$g_{1\gamma}\left(\alpha\right)=-\gamma e^{-\frac{\alpha^2}{\gamma}} \tag{5.62}$$

$$g_{2\gamma}\left(\alpha\right)=-\frac{\gamma}{1+\frac{\alpha^2}{\gamma}} \tag{5.63}$$

$$g_{3\gamma}\left(\alpha\right)=\gamma\ln\left(1+\frac{\alpha^2}{\gamma}\right) \tag{5.64}$$

$$g_{4\gamma}\left(\alpha\right)=\gamma\left|\alpha\right|-\gamma^2\ln\left(1+\frac{\left|\alpha\right|}{\gamma}\right) \tag{5.65}$$

The adaptive interaction function for Equation (5.65), i.e., fourth DA model, can be given by Equation (5.66):

$$h_{4\gamma}\left(\alpha\right)=\frac{1}{1+\frac{\left|\alpha\right|}{\gamma}} \tag{5.66}$$

Among all the aforementioned models given with this, AIF allows smoothening even at discontinuities to some extent. This overcomes the problem of boundless smoothing of standard regularization, when $\left|\alpha\right|\to\infty$ and abrupt end of smoothing as in the case of other regularizer model such as the line process model (LP).

5.8 CONVOLUTION BASED LOCAL INFORMATION IN FUZZY CLASSIFIERS

There are several methods for incorporating contextual information while processing remote sensing images. Contextual information can be incorporated from spatial domain called local information. Local information plays an important role to incorporate spatial and grey level information in an image. According to Krinidis and Chatzis (2010), the reasons behind incorporating local spatial and local grey level information are: (1) to sustain strength and noise insensitiveness, (2) to control the effect of neighborhood pixels, (3) to be independent from any fixed parameter, etc.

In the section (5.7), local information as spatial contextual information has been covered using smoothing and DA models. In this section, spatial contextual information has been included as three convolution based methods. In the past, many case studies related to adding local information as convolution methods with fuzzy based classifiers have been discussed. Ahmed et al. (2002) proposed FCM-S (fuzzy *c*-means with spatial constraints) in which objective function of FCM is modified in order to minimize the limitation of FCM by allowing the label of pixel to be influenced by labels of its neighbors. Chen and Zhang (2004) said inclusion of local information has several advantages over conventional fuzzy based classifiers such as: (1) if there is no previous knowledge of noise, (2) if it is free from any parameter selection, (3) if loss of image details is reduced, etc.

Ahmed et al. (2002) introduced FCM-S in which the objective function of FCM is modified in order to minimize the limitation of FCM by allowing the label of the pixel to be influenced by labels of its neighborhood.

Krinidis and Chatzis (2010) developed a robust fuzzy local information *c*-means (FLICM) clustering technique which incorporates local information to preserve image details and minimize the lack of robustness to noise and outliers. Using a spatial attraction model, a unique adaptive fuzzy logic local information *c*-means (ADFLICM) was developed by Zhang et al. (2017) for remotely sensed image classification in which they incorporate local spatial and grey level information to reduce edge blurring effect, and make it less sensitive to noise.

Fuzzy based classifiers work well on noise free images. Some of the fuzzy based classifiers are very sensitive to noise due to non-spatial contextual information being incorporated. To overcome the isolated noise based problem in fuzzy based classifiers, a preprocessing image smoothing step can be applied. Through smoothing filters such as low pass and median filters, important image details can be lost, especially boundaries or edges. However, there are fewer ways to control the trade-off between smoothing and clustering. In this context, many researchers have tried incorporating local spatial information with base fuzzy classification algorithms to improve the performance of classification, while removing noisy pixels.

The following section covers adding three types of local convolution methods, viz., fuzzy clustering with constraints, fuzzy local information, and adaptive fuzzy local information to the fuzzy based basic classifiers such as FCM, PCM, MPCM, IPCM, and NC.

5.8.1 Fuzzy c-Means with Constraints (FCM-S) Algorithm

Ahmed et al. (2002) introduced FCM-S in which the objective function of FCM was modified in order to reduce the limitation of FCM by allowing the label of a pixel to be influenced by labels of its neighborhood. The neighborhood effects as a regularizer and biases the solution toward piecewise homogenous labeling. The objective function for FCM-S can be given by Equation (5.67).

$$J_{\text{fcm}-s}(U,V) = \sum_{i=1}^{c}\sum_{k=1}^{N}\left(\mu_{ki}\right)^{m} D(x_k, v_i) + \frac{a}{N_R}\sum_{i=1}^{c}\sum_{k=1}^{N}\left(\mu_{ki}\right)^{m}\sum_{r\in N_k} D(x_r, v_i) \tag{5.67}$$

where x_r represents neighborhood of x_k, N_k represents the set of neighbors falling into a window around pixel x_k, a is the parameter for controlling the effect of neighbors, and N_R is the cardinality or convolution window size. By definition, each sample point x_k satisfies the constraints that $\sum_{i=1}^{c}\mu_{ki} = 1$.

The calculation of membership function from the above FCM-S objective function can be given by Equation (5.68):

$$\overline{\mu_{ki}} = \frac{\left[D(x_k, \overline{v_i}) + \frac{a}{N_R}\sum_{r\in N_k} D(x_r, \overline{v_i})\right]^{2}}{\sum_{j=1}^{c}\left[D(x_k, \overline{v_i}) + \frac{a}{N_R}\sum_{r\in N_k} D(x_r, \overline{v_i})\right]^{\frac{1}{m-1}}} \tag{5.68}$$

The second term in the numerator of Equation (5.68) is a neighbor average grey level value around x_k within a window. The images composed by all such neighbor average values around the image pixels are called mean filtered images.

5.8.2 Possibilistic c-Means with Constraints (PCM-S) Algorithm

PCM-S has been introduced as a novel term that enables the labeling of a pixel to be effected by labels of its neighbors (Singh and Kumar, 2019). The neighborhood effect acts as an allocator and drives the solution toward piecewise-homogeneous labeling. Mathematical formulation of PCM-S has been given in the following points:

1. Assign the number of class c, the value of degree of fuzziness $m > 1$, parameter "a," and convolution window size N_R.
2. Assign mean values for each class.
3. Compute the regularization parameter η_i.
4. Compute the membership matrix μ_{ki}, Equation (5.69):

$$\overline{\mu_{ki}} = \frac{1}{1+\left[\frac{D(x_k,\overline{v_i})+\frac{a}{N_R}\sum_{r\in N_k r\neq k} D(x_r,\overline{v_i})}{\eta_i}\right]^{\frac{1}{m-1}}} \tag{5.69}$$

5. Assign final class to each pixel.

The improved objective function of PCM-S is defined as follows in Equation (5.70):

$$J_{\text{pcm}-s}(U,V) = \sum_{i=1}^{c}\sum_{k=1}^{N}(\mu_{ki})^m D(x_k,v_i) + \sum_{i=1}^{c}\eta_i\sum_{k=1}^{N}(1-\mu_{ki})^m + \frac{a}{N_R}\sum_{i=1}^{c}\sum_{k=1}^{N}(\mu_{ki})^m\sum_{r\in N_k} D(x_r,v_i) \tag{5.70}$$

In Equation (5.69), x_r is a neighbor grey level value around x_k within a window. The image composed by all the neighbor average values around the image pixels forms a mean filtered image. The characteristics of PCM-S have been explained such as:

1. The neighborhood effect in PCM-S acts as regularizer and drives the solution toward piecewise-homogeneous labeling.
2. Incorporating spatial constraints in conventional PCM enhances insensitiveness to noise.

5.8.3 Fuzzy Local Information c-Means (FLICM) Algorithm

The FCM-S method has some disadvantages for assigning values to some of the parameters. Although the limitation of local spatial information enhances insensitiveness of FCM-S to noise somehow, still it is not robust to noise or outliers, especially in the absence of prior knowledge of the same. The crucial parameter a in the objective function of this method is used to balance between robustness to noise and effectiveness of preserving the details of the image. Generally, selection of this parameter is based on trial and error or experience. These are all applied on a static image and have to be computed in advance. To overcome these disadvantages, a new factor is required with some special characteristics as follows:

- It should be able to add local spatial with grey level information in fuzzy domain while preserving object boundaries.
- It should add the effect of the pixel in the spatial domain depending on their spatial space from a central pixel.

- It should use original image without requiring any preprocessing step and while preserving object boundary sharpness.
- It should be free of any parameter selection.

This novel fuzzy factor can be defined as in Equation (5.71):

$$G_{ki} = \sum_{j \in N_k k \neq j} \frac{1}{d_{kj}+1}\left(1-\mu_{ji}\right)^m D\left(x_j, v_i\right) \tag{5.71}$$

where d_{kj} is the spatial Euclidean distance between pixels k and j.

The advantage of factor G_{ki} is that it is independent of any variable that controls the balance between spatial and spectral information in the objective function. The control of this balance is due to parameters automatically generated and achieved by the definition of the fuzziness of each image pixel (both spatial and grey level). The Euclidean distance d_{kj} in the factor G_{ki} includes the effect of pixels within local window, to vary flexibly according to their distances from the central pixel. If window size is increased, then more local spatial information can be added at more generalization mode.

Thus, by using this factor G_{ki}, a more robust form of the FCM framework, viz. fuzzy local information c-means (FLICM) clustering algorithm, has been proposed. FLICM is capable to incorporate both the local spatial and grey level information. The objective function of this algorithm is defined as in Equation (5.72):

$$J_{\text{flicm}}(U,V) = \sum_{i=1}^{c} \sum_{k=1}^{N} \left(\mu_{ki}\right)^m D(x_k, v_i) + G_{ki} \tag{5.72}$$

The membership value for the above objective function can be computed as in Equation (5.73):

$$\overline{\mu_{ki}} = \frac{1}{\sum_{j=1}^{c} \left(\frac{D(k,i)+G_{ki}}{D\left(k,j\right)+G_{kj}} \right)^{\frac{1}{m-1}}} \tag{5.73}$$

5.8.4 Possibilistic Local Information c-Means (PLICM) Algorithm

Like the FLICM, the possibilistic local information c-means (PLICM) can be developed by adding fuzzy factor (G) for incorporating local information. The introduction of local information reduces the noise in the image. Like the FCM-S, the PCM-S is also not robust to noise while incorporating the local information. Again in this algorithm, the parameter a in the objective function also plays a crucial role for preserving the image details. Thus, in order to compensate these problems, the fuzzy factor (characteristic mentioned in the previous section) has been added to generate the PLICM. The mathematical formulation of the PLICM algorithm is as follows:

1. Assign the number of class c, the value of degree of fuzziness $m > 1$, and convolution window size N_R.
2. Assign mean values for each class.
3. Compute the regularization parameter η_k.
4. Compute the membership matrix μ_{ki}, Equation (5.74):

$$\overline{\mu_{ki}} = \frac{1}{1 + \left[\frac{D(x_k, \overline{v_i}) + G_{ki}}{\eta_i} \right]^{\frac{1}{m-1}}} \tag{5.74}$$

5. Assign final class to each pixel.

The improved objective function of PLICM is defined as follows in Equation (5.75):

$$J_{\text{plicm}}(U,V) = \sum_{i=1}^{c} \sum_{k=1}^{N} (\mu_{ki})^m D(x_k, v_i) + \sum_{i=1}^{c} \eta_i \sum_{k=1}^{N} (1 - \mu_{ki})^m + G_{ki} \tag{5.75}$$

By incorporating G_{ki}, PLICM is able to preserve more image details than PCM-S; however, it still has limitations while identifying the boundaries of the class as well as results over smoothed edges (Gong et al., 2012). Characteristics of PLICM are as mentioned below:

- It helps in preserving robustness and noise insensitiveness by incorporating local information in PCM.
- This algorithm is helpful in controlling the effect of the neighborhood data points or pixels depending on their distance from the central pixel.
- It is free of any parameter selection and independent of noise type.
- In PLICM, noise immunity and unwanted resistance property are completely dependent on the fuzzy factor G_{ki} as it is seen in the object function of this algorithm.
- G_{ki} is calculated automatically irrespective of whether it is manually set, and it can be possible without prior knowledge of noises.
- It preserves image details by incorporating G_{ki} with possibilistic based fuzzy spatial and grey level localities constraints (Krinidis and Chatzis, 2010).

The drawbacks of PLICM are as follows:

- In PLICM, G_{ki} cannot appropriately reproduce the damping extent of the neighbors.
- PLICM has limitations in identifying the edges, and it may result in over smoothed border.

5.8.5 Adaptive Fuzzy Logic Local Information *c*-Means (ADFLICM)

The adaptive fuzzy logic local information *c*-means (ADFLICM) approach with different fuzzy based classifiers covers spatial attraction between pixels. The attraction model has to be effective for characterizing the spatial correlation between pixels in the image. While adding the local spatial and grey level information, generalization of the attraction model is required. The spatial attraction model for two pixels j and k with respect to a given cluster i is the ratio of products of memberships with the spatial distance square between the two pixels as in Equation (5.76):

$$SA_{jk}(i) = \frac{\mu_{ji} \times \mu_{ki}}{D_{jk}^2} \tag{5.76}$$

where D_{jk} is a spatial distance between pixels j and k and is known as Chebyshev distance. This spatial attraction model gives birth to a unique local similarity measure S_{kr} and includes both spatial local and grey level information. This is well defined as in Equation (5.77):

$$S_{kr} = \begin{cases} SA_{kr}, k \neq r \\ 0, k = r \end{cases} \tag{5.77}$$

$$r \in 0 < (x_k - x_r)^2 + (y_k - y_r)^2 \leq 2^{L-1}$$

where k^{th} gives position of the pixel at the center of the local window whereas r^{th} represents neighborhood pixel, $(x_k - x_r)$ and $(y_k - y_r)$ denote the coordinates of pixel k and r respectively.

Based on this measure, the ADFLICM method is proposed for classification of remote sensing imagery. This was achieved by incorporating the local spatial information and grey level information in the objective function of conventional FCM classifier. The effect of this function is that results get smoothed toward piecewise-homogeneous classification, which reduces the edge blurring effect simultaneously. The objective function for this method is as in Equation (5.78):

$$J_{\text{adflicm}}(U,V) = \sum_{i=1}^{c}\sum_{k=1}^{N}(\mu_{ki})^m \left[D(x_k, v_i) + \frac{1}{N_R} \sum_{r \in N_k r \neq k} (1 - S_{kr}) D(x_r, v_i) \right] \tag{5.78}$$

By solving Equation (5.78), the membership function can be computed as Equation (5.79):

$$\overline{\mu_{ki}} = \frac{1}{\sum_{j=1}^{c} \left[\dfrac{D(x_k, \overline{v_i}) + \dfrac{1}{N_R} \sum_{r \in N_k r \neq k} (1 - S_{kr}) D(x_r, v_i)}{D(x_k, \overline{v_j}) + \dfrac{1}{N_R} \sum_{r \in N_k r \neq k} (1 - S_{kr}) D(x_r, v_j)} \right]^{\frac{1}{m-1}}} \tag{5.79}$$

5.8.6 Adaptive Possibilistic Local Information c-Means (ADPLICM) Algorithm

ADPLICM is an algorithm inspired by the advantages and disadvantages of PCM-S and PLICM. Like ADFLICM, this method also incorporates spatial attraction between pixels and local similarity measure in the image. Zhang et al. (2017) used the local similarity measure based on the pixel spatial attraction model in adaptive possibilistic local information *c*-means, which adaptively calculates the weighting factors for neighboring. Here an identical concept by including local information in characterizing the spatial correlation between pixels in the image (Mertens et al., 2006; Genyun et al., 2007) is introduced. Mathematical formulation of ADPLICM is mentioned as follows:

1. Assign the number of class c, the value of degree of fuzziness $m > 1$, and convolution window size N_R.
2. Assign mean values for each class.
3. Compute the regularization parameter η_i.
4. Compute the membership matrix μ_{ki} and local similarity measure S_{kr}, Equation (5.80):

$$\overline{\mu_{ki}} = \frac{1}{1+\left\{\dfrac{D(x_k,\overline{v_i})+\dfrac{1}{N_R}\sum\limits_{r\in N_k r\neq k}(1-S_{kr})D(x_r,v_i)}{\eta_i}\right\}^{\frac{1}{m-1}}} \tag{5.80}$$

5. Assign final class to each pixel.

The objective function of ADPLICM is defined as Equation (5.81):

$$\begin{aligned} J_{\text{adplicm}}(U,V) &= \sum_{i=1}^{c}\sum_{k=1}^{N}\left(\mu_{ki}\right)^m D(x_k,v_i)+\sum_{i=1}^{c}\eta_i\sum_{k=1}^{N}(1-\mu_{ki})^m \\ &+\sum_{i=1}^{c}\sum_{k=1}^{N}\left(\mu_{ki}\right)^m\frac{1}{N_R}\sum_{r\in N_k r\neq k}(1-S_{kr})D(x_r,v_i) \end{aligned} \tag{5.81}$$

The characteristics of ADPLICM are as follows:

- By incorporating the local spatial attraction model, the weight factors are calculated by both neighboring pixels and its center pixels concurrently.
- The ADPLICM is robust due to incorporation of the local similarity measures in possibilistic based *c*-means classifier.

- The noise acceptance and image detail conservation ability is completely reliant on the local spatial and grey level information from neighbors. It is controlled by S_{kr}.
- The local similarity measures S_{kr} balance their fuzzy membership values.
- S_{kr} suppresses the impacts from the noise of the neighbors.
- Incorporating local data in an assumed local window is independent of parameter selection.
- Preprocessing steps are not required, and classification can be accomplished directly on the original image to preserve the fine image details.
- It has strong impact in minimizing noisy pixels and reducing edge smoothing effects concurrently (Zhang et al., 2017).

5.8.7 Modified Possibilistic c-Means with Constraints (MPCM-S) Algorithm

The MPCM method discussed in the previous chapter has several disadvantages in the presence of noise in the image data. To overcome these, a new algorithm, modified possibilistic *c*-means with constraints (MPCM-S), has been introduced. In this algorithm all the properties of MPCM and convolution concepts are added to make the algorithm more immune to noise and outliers. This method has a novel term which makes the labeling of pixel affected by the label of neighbors. Mathematical formulation of MPCM-S is mentioned as follows:

1. Assign the number of class c, the value of degree of fuzziness $m > 1$, parameter a, and convolution window size N_R.
2. Assign mean values for each class.
3. Compute the regularization parameter η_i and noise minimizer λ_i.
4. Compute the membership matrix μ_{ki}, Equation (5.82):

$$\overline{\mu_{ki}} = \exp\left[-\frac{\left\{D(x_k, \overline{v_i}) + \frac{a}{N_R}\sum_{r \in N_k r \neq k} D(x_r, v_i)\right\}}{\eta_i}\right] \tag{5.82}$$

5. Assign final class to each pixel.

The modified objective function of MPCM-S is defined as in Equation (5.83):

$$\begin{aligned} J_{\text{mpcm-}s}(U,V) = {} & \sum_{i=1}^{c}\sum_{k=1}^{N}(\mu_{ki})^m D(x_k, v_i) + \sum_{i=1}^{c}\eta_i\sum_{k=1}^{N}(\lambda_i - \mu_{ki})^m \\ & + \frac{a}{N_R}\sum_{i=1}^{c}\sum_{k=1}^{N}(\mu_{ki})^m \sum_{r \in N_k r \neq k} D(x_r, v_i) \end{aligned} \tag{5.83}$$

where $\lambda_i = \mu_{ki} \log \mu_{ki}$.

5.8.8 Modified Possibilistic Local Information c-Means (MPLICM) Algorithm

To overcome the limitation of the MPCM method discussed in the previous section and to integrate the local spatial and grey level information for preserving image details as well as to control the effect of neighborhood pixels, the MPLICM method has been introduced. The mathematical formulation and objective function of MPLICM is represented as:

1. Assign the number of class c, the value of degree of fuzziness $m > 1$, and convolution window size N_R.
2. Assign mean values for each class.
3. Compute the regularization parameter η_i, noise minimizer λ_i and fuzzy factor G_{ki}.
4. Compute the membership matrix μ_{ki}, Equation (5.84):

$$\overline{\mu_{ki}} = \exp\left[-\frac{\left\{D(x_k,\overline{v_i}) + G_{ki}\right\}}{\eta_i}\right] \tag{5.84}$$

5. Assign final class to each pixel.

The improved objective function of PLICM is defined as in Equation (5.85):

$$J_{\text{mplicm}}(U,V) = \sum_{i=1}^{c}\sum_{k=1}^{N}(\mu_{ki})^m D(x_k, v_i) + \sum_{i=1}^{c}\eta_i \sum_{k=1}^{N}(\lambda_i - \mu_{ki})^m + G_{ki} \tag{5.85}$$

where $\lambda_i > 0$.

In PLICM and MPLICM both the noise acceptance and outliers confrontation property completely rely on the fuzzy factor G_{ki} as it is seen in the objective function of this algorithm, and G_{ki} is automatically determined.

5.8.9 Adaptive Modified Possibilistic Local Information c-Means (ADMPLICM) Algorithm

The ADMPLICM method has been introduced after incorporation of local similarity measures in MPCM. It is robust to noise due to this reason only. The mathematical formulation and objective function of this method is represented as follows:

1. Assign the number of class c, the value of degree of fuzziness $m > 1$, and convolution window size N_R.
2. Assign mean values for each class.

3. Compute the regularization parameter η_i noise minimizer λ_i and local similarity measure S_{kr} and μ_{ij} from Equation (5.86):

$$\overline{\mu_{ki}} = \exp\left[-\frac{\left\{D(x_k, \overline{v_i}) + \frac{1}{N_R}\sum_{r\in N_k r\neq k}(1-S_{kr})D(x_r, v_i)\right\}}{\eta_i}\right] \tag{5.86}$$

4. Assign final class to each pixel.

The objective function of ADPLICM is defined as follows in Equation (5.87):

$$J_{\text{admplicm}}(U,V) = \sum_{i=1}^{c}\sum_{k=1}^{N}(\mu_{ki})^m D(x_k, v_i) + \sum_{i=1}^{c}\eta_i\sum_{k=1}^{N}(\lambda_i - \mu_{ki})^m + \sum_{i=1}^{c}\sum_{k=1}^{N}(\mu_{ki})^m \frac{1}{N_R}\sum_{r\in N_k r\neq k}(1-S_{kr})D(x_r, v_i) \tag{5.87}$$

where $\lambda_i > 0$.

We need to add a good assessment of methods to be discussed in terms of computation and accuracies, summary of methods made, and where each hybrid system is beneficial.

5.9 SUMMARY

In this chapter, hybrid classifiers covered with fuzzy based base classifiers and adding MRF or local neighbor information have been discussed. As in these classifiers, only Euclidean distance has been used; in this chapter various single distance, composite distance, as well as kernels we have induced. Later contextual information through MRF as well as local convolution based approach have been described. The next chapter will focus in on fuzzy based classifiers capable of classifying a single class of interest.

BIBLIOGRAPHY

Abello, J., Pardalos, P.M. and Resende, M.G.C., 2013. *Handbook of Massive Data Sets*, vol. 4. New York: Springer-Verlag.

Ahmed, M.N., Yamany, S.M., Mohamed, N., Farag, A.A. and Moriarty, T., 2002. A modified fuzzy *c*-means algorithm for bias field estimation and segmentation of MRI data. *IEEE Transactions on Medical Imaging*, 21 (3), 193–199. https://doi.org/10.1109/42.996338

Bezdek, J.C., 1981. *Pattern Recognition with Fuzzy Objective Function Algorithms*. New York: Plenum Press.

Bezdek, J.C., Ehrlich, R. and Full, W., 1984. FCM: The fuzzy *c*-means clustering algorithm. *Computers & Geosciences*, 10, 191–203.

Bloom, S.A., 1981. Similarity indices in community studies: potential pitfalls. *Marine Ecology Progress Series*, 5 (2), 125–128.

Boser, H., Guyon I.M. and Vapnik, V.N., 1992. A training algorithm for optimal margin classifiers. In: *Proceedings of 5th Annual ACM Workshop on Computational Learning Theory*, Pittsburgh, PA, pp. 144–152.

Camps-Valls, G. and Bruzzone, L., 2009. *Kernel Methods for Remote Sensing Data Analysis*, Hoboken: Wiley, pp. 25–45.

Camps-valls, G., Gomez-chova, L., Muñoz-marí, J., Vila-francés, J. and Calpe-maravilla, J., 2006. Composite kernels for hyperspectral image classification. *IEEE GeoScience and Remote Sensing Letters*, 3 (1), 93–97.

Cantrell, C.D., 2000. *Modern Mathematical Methods for Physicists and Engineers*, Cambridge, UK: Cambridge University Press.

Chang, C.I., 2000. An information-theoretic approach to spectral variability, similarity, and discrimination for hyperspectral image analysis. *IEEE Transactions on Information Theory*, 46 (5), 1927–1932. https://doi.org/10.1109/18.857802

Chattopadhyay, S., Pratihar, D.K. and Sarkar, S.C.D., 2011. A comparative study of fuzzy *c*-means algorithm and entropy-based fuzzy clustering algorithms. *Computing and Informatics*, 30, 701–720.

Chen, S. and Zhang, D., 2004. Robust image segmentation using FCM with spatial constraints based on new kernel-induced distance measure. *IEEE Transactions on Systems, Man, and Cybernetics, Part B: Cybernetics*, 34 (4), 1907–1916. https://doi.org/10.1109/TSMCB.2004.831165.

Cortes C. and Vapnik V., 1995. Support-vector networks. *Machine Learning*, 20 (3), 273–297. https://doi.org/10.1007/BF00994018.

Duda, R.O., Hart, P.E. and Stork, D.G., 2001. *Pattern Classification*. New York: Wiley.

Dunn, J.C., 1973. A fuzzy relative of the ISODATA process and its use in detecting compact well-separated clusters. *Cybernetics and Systems*, 3, 32–57.

Emran, S.M., and Ye, N., 2002. Robustness of chi-square and canberra distance metrics for computer intrusion detection. *Quality and Reliability Engineering International*, 18 (1), 19–28.

Feyerabend, P.K., 2015. *Physics and Philosophy: Volume 4: Philosophical Papers*. Cambridge, UK: Cambridge University Press.

Francis, T.S., 2000, *Entropy and Information Optics. Optical Science and Engineering*. 1st Edition. CRC Press.

Geman, S. and Geman, D., 1984. Stochastic relaxation, gibbs distributions, and the bayesian restoration of images. *IEEE Transactions on Pattern Analysis and Machine Intelligence, PAMI*, 6 (6), 721–741. https://doi.org/10.1109/TPAMI.1984.4767596.

Gong, M., Zhao, J., Liu, J., Miao, Q. and Jiao, L., 2012. Change detection in synthetic aperture radar images based on deep neural networks. *IEEE Transactions on Neural Networks and Learning Systems*, 27 (1), 125–138. https://doi.org/10.1109/TNNLS.2015.2435783.

Goshtasby, A.A., 2012. Similarity and dissimilarity measures. In: Goshtasby, A.A. (ed.), *Image Registration*. pp. 7–66. London: Springer.

Hasnat, A., Halder, S., Bhattacharjee, D., Nasipuri, M. and Basu, D. K., 2013. Comparative study of distance metrics for finding skin color similarity of two color facial images. *National Conference on Advancement of Computing in Engineering Research (ACER 13), West Bengal, India.*

Jurman, G., Riccadonna, S., Visintainer, R. and Furlanello, C., 2009. Canberra distance on ranked lists. In: *Proceedings of Advances in Ranking NIPS 09 Workshop*, pp. 22–27.

Krinidis, S. and Chatzis, V., 2010. A robust fuzzy local information *c*-means clustering algorithm. *IEEE Transactions on Image Processing: A Publication of the IEEE Signal Processing Society*, 19 (5), 1328–1337. https://doi.org/10.1109/TIP.2010.2040763.

Kumar, A., 2007. *Investigation in Sub-Pixel Classification Approaches for Land Use and Land Cover Mapping*. PhD Thesis, I.I.T. Roorkee, India.

Kumar, A. and Dadhwal, V.K., 2010. Entropy-based fuzzy classification parameter optimization using uncertainty variation across spatial resolution. *Journal of the Indian Society of Remote Sensing*, 38 (2), 179–192.

Lance, G.N. and Williams, W.T., 1966. Computer programs for hierarchical polythetic classification ("similarity analyses"). *The Computer Journal*, 9 (1), 60–64.

Lance, G.N., and Williams, W.T., 1967. Mixed-data classificatory programs I—Agglomerative systems. *Australian Computer Journal*, 1 (1), 15–20.

Li, S.Z., 2009. *Markov Random Fields Modeling in Image Analysis*. 3rd Edition. London: Springer. *e-book*.

Li, R.P. and Mukaidono, M., 1999. Gaussian clustering method based on maximum-fuzzy entropy interpretation. *Fuzzy Sets and Systems*, 102, 253–258.

Melgani, F. and Serpico, S.B., 2003. A Markov random field approach to spatio-temporal contextual image classification. *IEEE Transactions on Geoscience and Remote Sensing*, 41 (11), 2478–2487.

Mercier, G. and Lennon, M., 2003. Support vector machines for hyperspectral image classification with spectral-based kernels. *IGARSS*, IEEE, Toulouse, France, pp. 288–290.

Mertens, K.C., de Baets, B., Verbeke, L.P.C. and de Wulf, R.R., 2006. A sub-pixel mapping algorithm based on sub-pixel/pixel spatial attraction models. *International Journal of Remote Sensing*, 27 (15), 3293–3310. https://doi.org/10.1080/01431160500497127.

Metropolis, N., Rosenbluth, A.W., Rosenbluth, M.N., Teller, A.H. and Teller, E., 1953. Equation of state calculations by fast computing machines. *The Journal of Chemical Physics*, 21, 1087.

Miyamoto, S. and Mukaidono, M., June 25–30, 1997. Fuzzy *c*-means as a regularization and maximum entropy approach. In: *Proceeding of the 7th International Fuzzy Systems Association World Congress (IFSA 1997)*, vol. II, pp. 86–92. Prague, Czech.

Miyamoto, S., Ichihashi, H. and Honda, K., 2008. *Algorithms for Fuzzy Clustering, Studies in Fuzziness and Soft Computing*, vol. 229, pp. 65–66. Springer.

Naresh Kumar, M., Seshasai, M.V.R., Vara Prasad, K.S., Kamala, V., Ramana, K.V., Dwivedi, R.S. and Roy, P.S., 2011. A New Hybrid Spectral Similarity Measure for Discrimination Among Vignaspecies. *International Journal of Remote Sensing*, 32 (14), 4041–4053. https://doi.org/10.1080/01431161.2010.484431.

Refaat, M.M. and Farag, A.A., 2004. *Mean Field Theory for density Estimation Using Support Vector Machines, Computer Vision and Image Processing Laboratory*. Louisville, KY: University of Louisville, 40292.

Scollar, I., Weidner, B. and Huang, T.S., 1984. Image enhancement using the median and the interquartile distance. *Computer Vision, Graphics, and Image Processing*, 25 (2), 236–251.

Shannon, C.E., 1948. A mathematical theory of communication. *American Telephone and Telegraph Technology Journal*, 27, 379–423.

Shannon, C., 1951. Prediction and entropy of printed English. *Bell Systems Technical Journal*, 30, 50–64.

Singh, A. and Kumar, A., February 20, 2019. Fuzzy based approach to incorporate spatial constraints in possibilistic *c*-means algorithm for remotely sensed imagery. In: *International Conference on Sustainable Computing in Science, Technology & Management*, Jaipur, p. 5. https://doi.org/http://dx.doi.org/10.2139/ssrn.3354465.

Solberg, A.H. S., Taxt, T. and Jain, A.K.,1996. A Markov random field model for classification of multisource satellite imagery. *IEEE Transactions on Geoscience and Remote Sensing*, 34 (1), 100–113.

Sun, G., Liu, Q., Liu, Q., Ji, C. and Li, X., 2007. A novel approach for edge detection based on the theory of electrostatic field. In: *2007 International Symposium on Intelligent Signal Processing and Communications Systems, ISPACS 2007—Proceedings*, 40, 260–263. https://doi.org/10.1109/ISPACS.2007.4445873.

Theodoridis, S. and Koutroumbas, K., 2006. Clustering: Basic concepts. *Pattern Recognition*, 3, 483–516.

Tihonov, A.N. and Arsenin, V.Y., 1977. Solutions of Ill-posed problems. New York: Wiley.

Tso, B. and Mather, P.M., 2009. *Classification Methods for Remotely Sensed Data*, Boca Raton: CRC Press, pp. 56–69.

Van Der Heijden, F., Duin, R., De Ridder, D. and Tax, D.M.J., 2005. *Classification, Parameter Estimation and State Estimation: An Engineering Approach Using MATLAB*. Chichester, UK: John Wiley & Sons.

Van der Meer, F., 2006. The effectiveness of spectral similarity measures for the analysis of hyperspectral imagery. *International Journal of Applied Earth Observation and Geoinformation*, 8(1), 3–17. https://doi.org/10.1016/j.jag.2005.06.001.

Vapnik, V., 1995. *The Nature of Statistical Learning Theory*. New York: Springer-Verlag.

Vapnik, V., 1998. *Statistical Learning Theory*. New York: John Wiley and Sons.

Vapnik, V.N., 1999. An overview of statistical learning theory. *IEEE Transactions of Neural Networks*, 10, 988–999.

Vapnik, V.N., 2000. *The Nature of Statistical Learning Theory*. 2nd Edition. New York: Springer-Verlag.

Vassiliadis, S., Hakkennes, E.A., Wong, J. and Pechanek, G.G., 1998. The sum-absolute-difference motion estimation accelerator. In: *Euromicro Conference, 1998. Proceedings, 24th*, vol. 2, pp. 559–566.

Von Luxburg, U., 2004. Statistical learning with similarity and dissimilarity functions. Available from http://www.tml.cs.uni-tuebingen.de/team/luxburg/publications/Luxburg04_diss.pdf.

Yao, J., Dash, M., Tan, S.T. and Liu, H., 2000. Entropy-based fuzzy clustering and fuzzy modeling. *Fuzzy Sets and System*, 113, 381–388.

Ye, J., 2011. Multicriteria decision-making method based on a cosine similarity measure between trapezoidal fuzzy numbers. *International Journal of Engineering, Science and Technology*, 3 (1), 272–278.

Zhang, M., Therneau, T., McKenzie, M.A., Li, P. and Yang, P., 2008. A fuzzy *c*-means algorithm using a correlation metrics and gene ontology. In: *19th International Conference on Pattern Recognition (ICPR 2008)*, December 8–11, 2008, IEEE Computer Society, Tampa, FL.

Zhang, H., Wang, Q., Shi, W. and Hao, M., 2017. A novel adaptive fuzzy local information *c*-means clustering algorithm for remotely sensed imagery classification. *IEEE Transactions on Geoscience and Remote Sensing*, 55 (9), 5057–5068. https://doi.org/10.1109/TGRS.2017.2702061.

6 Fuzzy Classifiers for Temporal Data Processing

6.1 INTRODUCTION

In the budding stage of use of remote sensing technology, emphasis was generally placed on the use of single date multispectral data for identification of land cover. The single date image, however, may have some inherent problems because the DN values of objects in the image do not have a fixed value. Further it may also be possible to have similar DN values for two spectrally dissimilar objects due to background or atmospheric conditions. Therefore, single date remote sensing imagery for the extraction of a specific class is generally not recommended. Due to the technological development in recent years, the use of temporal, multispectral data for identification of a specific class is recommended (Dadhwal et al., 2002). In the case of temporal data, the spectral response of a land class proportion can be recognized due to its unique temporal characteristics.

Classification of vegetation classes beyond the third level, which in other words can be related to discrimination between various agriculture crops and varieties, can be done using the growth pattern of that crop or vegetation. While mapping a specific crop using remote sensing techniques, the information of other crops present in that area or nearby is also important. Otherwise the overlapping of spectral responses of different crops makes the mapping of a specific crop using single date imagery a real challenge. Analysis of specific crop in temporal domain can, however, provide a solution for discriminating it from other crops and vegetation classes. Thus, the growth pattern of a crop may act as a discriminating factor indeed. The availability of temporal data is necessary for continuous monitoring corresponding to the phenological changes of crops and vegetation.

The mixed pixel is another problem while preparing accurate crop maps. For accurate mapping of the crop, contribution due to the fractional part of the pixel as well as full crop pixel need to be included. The proportional contribution or un-mixing of classes can be handled using techniques like linear mixture model (LMM), fuzzy classification, neural network, etc. As a result, the precise area estimation of a land cover can easily be found for both low and high spatial resolution imageries (Dadhwal et al., 2002).

The crop pattern in India, like some other parts of the world, is heterogeneous in nature. This signifies that in most cases, crops are generally grown in a small patch of land. Thus, in most cases, the spectral responses of crops may overlap in single date imagery, hence there is need of temporal images for identifying the single land cover from remote sensing images (Wardlow et al., 2007; Masialeti et al., 2010). This overlap can also be due to cultural practices adopted by the farmers such as planting date, physiological status of crop, etc. Several studies show that the time series analysis of crops is a suitable technique for identifying a particular crop among various crops and vegetation classes present in the image.

Another important hurdle for temporal analysis is availability of clouds-and haze-free scenes throughout the season or in case of crops during the growth period. Uniqueness of crop or growth stages or phenological changes can be triggering factors for identifying them independently. The requirement of noise free temporal images is rarely met due to poor atmospheric conditions as well as constraints of temporal coverage of satellite. Thus, there remain gaps in the data sampling while doing temporal analysis (Steven et al., 2003). However, these gaps can be filled by a multi-sensor based analysis.

6.2 TEMPORAL INDICES APPROACH

Spectral indices are helpful for emphasizing various land cover present in the image. There are different indices for different types of land cover such as vegetation, urban, water and snow, etc. These spectral indices in temporal domain may be very helpful for identification of that land cover.

The green vegetation canopies have a very distinctive interaction with the energy in the visible and infrared region of the electromagnetic spectrum. Vegetation classes have a distinctive nature of interaction with the electromagnetic energy in visible and infrared bands. In the presence of chlorophyll in vegetation, the red and blue bands of the visible region of the electromagnetic spectrum absorb energy strongly and hence the vegetation classes generally appear green. Further, in the near infrared band the energy is scattered strongly by the vegetation class due to internal structure and moist leaves and hence results in a very high reflectance. The aforementioned spectral characteristics of green vegetation are key factors for the evolution of vegetation indices. Vegetation index (VI) is a dimensionless and radiometric measure of green vegetation (Ghosh, 2013).

The vegetation indices are generally formulated by the combination of visible and infrared spectral values. They are carried out by the arithmetic operations and yield a single value, which indicates the amount of vegetation within that pixel. Vegetation indices computed from satellite images can give a good indication of the presence of vegetation in many cases, as well as minimizing the effect of shadow (Campbell, 1987). A time series analysis of vegetation indices for a season may help for next level of vegetation classification such as crop or forest. Time series normalized difference vegetation index (NDVI) data has been popularly used in continuously monitoring land cover characteristics (Tingting and Chuang, 2010; Pringle et al., 2012) and vegetation phenology (Sakamoto et al., 2005).

While using multispectral remote sensing directly for vegetation class identification, dimensionality is another aspect to keep in mind. This is due to the fact that the increasing number of bands in a multispectral image increases dimensionality. In addition to this, to observe the temporal change in a land cover type (say, crop or vegetation), the dimensionality further increases manifold.

On the other hand, the vegetation index for multispectral image not only reduces the dimensionality but also enhances the vegetation feature present in the image. Therefore, temporal spectral indices are the first choice for identification of a land cover uniquely. A variety of spectral indices exist and can be used in temporal domain for identifying various land cover (Table 6.1).

Generally, the spectral indices (especially NDVI derived) are based on the concept of maximum and minimum reflectance. Therefore, another type of vegetation index has been used by the researchers (Sengar et al., 2012; Upadhyay et al., 2012) which generally depends upon maximum and minimum spectral value of

TABLE 6.1
List of Some Commonly Used Spectral Indices

1. **Simple Ratio (SR)**

$$\mathrm{SR} = \frac{\rho_{\mathrm{NIR}}}{\rho_R}$$

where ρ_{NIR} is the reflectance at NIR band and ρ_R is the reflectance at red band.

Birth and McVey (1968)

2. **Normalized Difference Vegetation Index (NDVI)**

$$\mathrm{NDVI} = \frac{\rho_{\mathrm{NIR}} - \rho_R}{\rho_{\mathrm{NIR}} + \rho_R}$$

Rouse et al. (1973)

3. **Soil Adjusted Vegetation Index (SAVI)**

$$\mathrm{SAVI} = \frac{(\rho_{\mathrm{NIR}} - \rho_R)(1+L)}{\rho_{\mathrm{NIR}} + \rho_R + L}$$

where L is a canopy background adjustment factor and generally has a value of 0.5.

Running et al. (1994); Huete et al. (1994)

4. **Enhanced Vegetation Index (EVI)**

$$EVI = \left[\frac{\rho_{\mathrm{NIR}} - \rho_R}{\rho_{\mathrm{NIR}} + C_1\rho_R - C_2\rho_B + L}\right]G$$

The coefficients C_1, C_2, and L have been empirically determined as 6.0, 7.5, and 1.0, respectively, with G having a value of 2.5.

Huete et al. (1994, 1997)

5. **Enhanced Vegetation Index 2 (EVI2)**

$$EVI2 = \frac{2.5 \times (\rho_{\mathrm{NIR}} - \rho_R)}{\rho_{\mathrm{NIR}} + 2.4 \times \rho_R + 1}$$

Jiang et al. (2008)

6. **Transformed Vegetation Index (TVI)**

$$TVI = \sqrt{\left(\frac{\rho_{\mathrm{NIR}} - \rho_R}{\rho_{\mathrm{NIR}} + \rho_R}\right) + 0.5}$$

Deering et al. (1975)

7. **Normalized Burn Ratio (NBR)**

$$\mathrm{NBR} = \frac{\rho_{\mathrm{NIR}} - \rho_{\mathrm{MIR}}}{\rho_{\mathrm{NIR}} + \rho_{\mathrm{MIR}}}$$

Key and Benson (2006)

8. **Normalized Difference Water Index (NDWI)**

$$\mathrm{NDWI} = \frac{\rho_{\mathrm{NIR}} - \rho_{\mathrm{SWIR}}}{\rho_{\mathrm{NIR}} + \rho_{\mathrm{SWIR}}}$$

NIR = 0.86 μm and SWIR = 1.24 μm

Gao (1996)

9. **Normalized Difference Snow Index (NDSI)**

$$\mathrm{NDSI} = \frac{\rho_{\mathrm{green}} - \rho_{\mathrm{SWIR}}}{\rho_{\mathrm{green}} + \rho_{\mathrm{SWIR}}}$$

Green = 0.5–0.6 μm and SWIR = 1.55–1.75 μm

Hall et al. (1995)

10. **Atmospherically Resistant Vegetation Index (ARVI)**

$$\mathrm{ARVI} = \frac{\rho_{\mathrm{NIR}} - \rho_{rb}}{\rho_{\mathrm{NIR}} + \rho_{rb}}$$

$$\rho_{rb} = \rho_{\mathrm{red}} - \gamma(\rho_{\mathrm{blue}} - \rho_{\mathrm{red}})$$

γ mainly depends on the type of aerosol size.

Kaufman and Tanre (1992)

a particular class. For example, if a user is interested to identify a particular land cover (say, crop), then his/her interest will be to observe the maximum and minimum reflecting irrespective of satellite sensor wavelength. Such type of index is known as a class based sensor independent index (CBSI) and can be defined as in Equation (6.1):

$$\text{CBSI} = \frac{\rho_{\text{max}} - \rho_{\text{min}}}{\rho_{\text{max}} + \rho_{\text{min}}} \tag{6.1}$$

6.3 FEATURE SELECTION METHODS

Feature selection is required while using the multi-temporal data to reduce the number of unwanted scenes. It is a method to segregate the combination of useful bands or scenes for identifying a particular land cover using the temporal multispectral data (Bruzzone and Serpico, 2000). The increasing number of input features increases the computational requirement as well as cost. The feature selection method is used for generating the suitable combinations of datasets for achieving the maximum accuracy with low cost and less labor (Tso and Mather, 2009). The feature selection is different from the feature extraction method (e.g., principal component analysis, decision tree) in the sense that it uses the separability distances in the input feature space to derive the best sub feature dimensions. On the other hand, feature extraction compresses the information available in the original feature space with a drawback of losing the physical significance of features (Bruzzone and Serpico, 2000).

There are few methods, such as city block distance, Euclidean distance, angular separation, normalized city block distance, divergence, transformed divergence (TD), Bhattacharyya's distance, and Jeffreys-Matusita (JM) to measure the separability distances (Ghosh, 2013). Out of all these methods, the last four methods are commonly used for feature selection by determining the separability measure of remote sensing data, which is further used as an important criterion for feature selection. The separability measure tries to select the best number of bands to be used out of the given dataset. Suppose that there are n numbers of bands in a given dataset, and an analyst is interested in finding the best q number of bands, then the number of band combination C to be examined at a time can be expressed as in Equation (6.2) (Jensen, 1996; Ghosh, 2013):

$$C\left(\frac{n}{q}\right) = \frac{n!}{q!(n-q)!} \tag{6.2}$$

Transformed divergence (TD) (Swain and Davis, 1978) and Jeffreys-Matusita (JM) distance (Swain and Davis, 1978) separability approach are commonly used separability measures. Transformed divergence (TD) can be expressed as in Equation (6.3):

$$TD_{ij} = 2{,}000\left\{1 - \exp\left(\frac{-D_{ij}}{8}\right)\right\} \tag{6.3}$$

where:
i and j are two signatures (classes) being compared and
D_{ij} is the divergence.

The divergence D_{ij} can be calculated by the following Equation (6.4).

$$D_{ij} = \frac{1}{2} tr\left(\left(C_i - C_j\right)\left(C_i^{-1} - C_j^{-1}\right)\right) + \frac{1}{2} tr\left(\left(C_i^{-1} - C_j^{-1}\right)\left(u_i - u_j\right)\left(u_i - u_j\right)^T\right) \quad (6.4)$$

where:
C_i is the covariance matrix of class i,
u_i is the mean vector of class i,
tr is the trace function, and
T is the transpose function of the matrix.

The TD values range from 0 to 2,000. According to Jensen (1996), for TD value greater than 1,900, there is no overlapping between classes thereby having good separation. If it lies between 1,700 and 1,900, the separation is fairly good, whereas for less than 1,700, separation is poor. On the other hand, the JM distance ranges from 0 to 1,414. The JM distance can be expressed as in Equations (6.5) and (6.6):

$$JM_{ij} = \sqrt{2\left(1 - \exp(-\alpha)\right)} \quad (6.5)$$

where:

$$\alpha = \frac{1}{8}\left(u_i - u_j\right)^T \left(\frac{C_i - C_j}{2}\right)^{-1} \left(u_i - u_j\right) + \frac{1}{2} \ln\left(\frac{\left|C_i + C_j\right|/2}{\sqrt{\left|C_i\right| \times \left|C_j\right|}}\right) \quad (6.6)$$

where:
i and j are two signatures (classes) being compared,
C_i is the covariance matrix of class i,
u_i is the mean vector of class i, and
$\left|C_i\right|$ is the determinant of matrix C_i.

6.4 SOME CASE STUDIES FOR TEMPORAL DATA ANALYSIS

The dynamic nature of a few land cover over a period is a triggering factor to utilize the temporal data for mapping a specific land cover class. Let us take an example of a vegetation class in which the phenological changes over a period of time will be a unique factor to discriminate it from the other classes. In many studies, the temporal remote sensing data have been used for different applications such as estimation of forest biomass (Powell et al., 2010), flood study (Sakamoto et al., 2007), forest fires (Goetz et al., 2006; Morton et al., 2011), forest mapping (Hilker et al., 2009) and landscape changes (Millward et al., 2006). The multi-temporal MODIS satellite data have been used for the crop studies (Xiao et al., 2006; Wardlow et al., 2007; Wardlow and Egbert, 2008, 2010; Pan et al., 2012, Upadhyay et al., 2016).

Due to its coarse spatial resolution, the MODIS dataset is suitable for crop mapping at local to regional scales. The MODIS time series data at 250 m spatial resolution has been used by researchers for identification of forest area estimation (Maselli, 2011), tropical forest phenology (Pennec et al., 2011), gross primary production (Schubert et al., 2012), and identification of cropping activity (Pringle et al., 2012). Some of the case studies for identification of land cover are discussed in the following.

1. Wang and Tenhunen (2004) used the multi-temporal NDVI data from NOAA-AVHRR for different vegetation mapping in northeastern China for the year 1997. Supervised minimum distance and unsupervised *k*-means classification methods have been applied on the temporal NDVI data and its phenology based derived matrices such as maximum, mean, threshold, amplitude, total length of growing season, fraction of growing season during green up, rate of green up, rate of senescence, etc. The overall accuracy for NDVI temporal profile for unsupervised *k*-means and supervised minimum distance were 52% and 50%, respectively. The overall accuracy for NDVI derived matrix was below 50%. Thus, classifications based on the NDVI temporal profile were better than those with the derived matrices.
2. Blaes et al. (2005) used the three optical images along with a number of time series synthetic aperture radar (SAR) images for crop identification. The idea of using the SAR images along with the optical was to overcome the problem due to cloud cover conditions and to guarantee necessary temporal frequency throughout the growing season. The classification was performed by the different combination of optical or SAR imagery independently. The main focus was to study the effect due to inclusion of SAR images on the optical images. It was found that the classification accuracy increased by at least 5% when SAR images were combined with the optical images alone.
3. Xiao et al. (2006) mapped paddy rice fields in south Asia and southeast Asia using the multi-temporal MODIS images. A MOD09A1 product with a spatial resolution of 500 m and composite period of 8 days was used for the study. Out of 46 tiles of MOD09A1 for the year 2002, only 23 were selected for the study. The paddy rice mapping algorithm that uses the time series of MODIS derived vegetation indices was used for the analysis. The resultant maps were compared with the agricultural statistical data at national and sub national levels. The outputs for the MODIS rice algorithm were similar to the database derived from the census statistics.
4. Wardlow et al. (2007) investigated that the MODIS 250 m 12-month time series (January–December 2001) VI data. It was found that the data had the sufficient spatial, spectral, and temporal resolution to discriminate the crop types for Kansas in U.S. central Great Plains. Climatic and management

practice variation was also detected for the crop class of study in the time series data. The phenological profiles which were spectrally and temporally different for different crops had been observed. A similar cropping pattern was observed for the MODIS 250 m and Landsat ETM+ 30 m imagery. It was found that MODIS 250 m is an appropriate scale to measure the general crop mapping pattern for the U.S. central Great Plains with the field with size 32.4 hectare or larger. The possibility of sub pixel un-mixing was also carried out to estimate the proportion of specific land cover class.

5. Wardlow and Egbert (2008) evaluated the applicability of time series MODIS 250 m normalized difference vegetation index (NDVI) data spanning from March 22 to November 1, 2001, for large-area crop-related LULC mapping over the U.S. central Great Plains. A hierarchical crop mapping protocol was applied to a decision tree classifier using multi-temporal NDVI data collected over the crop growing season for the state of Kansas. Classification accuracies for the time series MODIS NDVI derived crop maps were greater than 80%. Overall accuracies ranged from 94% to 84% for the general crop map and summer crop map, respectively
6. Tingting and Chuang (2010) used the NDVI, NDWI, and normalized difference soil index (NDSoI) based time series spectral indices (12 periods out of the time series dataset from April 2007 to October 2007) for identifying the rice crop in the Chao Phraya Basin of Thailand. The first principal component corresponding to each of the three MODIS time series was combined to create a new dataset. A linear spectral un-mixing was then applied to this merged data to create another data set. Thereafter, using the composition of NDVI, NDWI, and NDSoI values in each pixel, agricultural crop land has been separated into upland and paddy fields in Thailand by using support vector machine (SVM).
7. Wardlow and Egbert (2010) performed a comparative study between Terra MODIS-250 m NDVI and EVI data, acquired from March 22 to September 30, 2001, for different crop mapping. The study was carried out under the assumption that EVI is more sensitive for crop mapping studies. The study was carried for the U.S. central Great Plains for general crop types, summer crop types, and irrigated and non-irrigated crops. It was observed that the NDVI and EVI produced equivalent crop classification with a subtle difference in their multi-temporal behavior. The overall and class specific classification accuracies were greater than 85% for both NDVI and EVI. The variation in the classification accuracy between the maps was of the order of 3%, and their pixel level agreement was greater than 90%. Since this study was performed for a small geographical area and for a single season, the applicability of this study can be verified after investigating its inter-annual climatic behavior and performing it for other major agricultural regions of the world.
8. Potgieter et al. (2010) provided the early season information on winter crop (wheat, barley, and chickpea) area estimates using the multi-temporal MODIS 250 m EVI data acquired for the period 2003–2004 for a study area in Queensland, Australia. This study was aimed to fulfill the requirement

of the early estimates of net crop production before the harvest, which is useful for many applications such as the grain industry, disaster relief, and drought declaration. The unsupervised *k*-means algorithm was used for classification. The study shows that the multi-temporal remote sensing approach could be used for the early season crop area prediction, at least 1–2 months before the harvesting date.

9. Atzberger and Eilers (2011) used a time series data consisting of the 10-day maximum value composite images from the SPOT VGT for monitoring the vegetation activity and phenology in South America at a spatial resolution of 1 km, from April 1998 to December 2008. The Whittaker smoother (WS) filter was applied on the time series data to handle the missing data, filter noise, and construct high quality NDVI time series. The geostatistical variogram technique was applied to reveal signal to noise ratio (SNR) WS filtered images. It was found that the filtered time series had the potential to distinguish between various plant functional types, as well as a key for various phenological markers. Thus, it was concluded that the time series datasets have great potential for vegetation and environmental related studies.
10. Alcantara et al. (2012) used the multi-temporal Terra and Aqua MODIS 250 m NDVI data for mapping of abandoned agricultural fields in Eastern Europe. TIMESAT software-derived phenological parameters were used as input parameter for SVM based classification of these agriculture fields. For classification, an overall accuracy of 65% for growing season has been achieved. Although the multi-year MODIS NDVI data does not increase the classification accuracy, but by using phenology matrices the accuracy has been increased by 8%.
11. Gonçalves et al. (2012) applied univariate and multi-variate statistical forecasting models to compute the water requirement satisfaction index (WRSI) and NDVI from AVHRR time series satellite images to monitor the sugarcane fields in Brazil from April 2001 to March 2008. Although both the models successfully predicted the NDVI values, the accuracy of the univariate model was higher than the multi-variate model. The average relative prediction error in case of univariate and multi-variate models were 5.6% and 13.4%, respectively. The forecasting of WRSI has given higher prediction error of order 49.7% and 47% for univariate and multivariate respectively. This is due to the fact that the WRSI values vary frequently throughout the season. An autocorrelation between these two indices has shown a time lag of one month for NDVI, which means the NDVI values change after approximately one month of climate change occurs.
12. Pan et al. (2012) proposed a crop proportion phenology index (CPPI) to estimate the winter wheat crop area up to the sub pixel level by using MODIS EVI time series for two agricultural regions in China. The phenological variables from October 2006 to June 2007 were used

as an input to calculate CPPI for both the study areas. CPPI has been estimated by fitting either the linear or non-linear regression models on the phenological variables. The inversion model has been used to calculate the regression coefficients using the training samples. The utility of the index was tested on two experimental areas in China. It was found that the CPPI performed well in fractional crop area predictions, with RMSE ranging roughly from 15% in the individual pixel to 5% above 6.25 km^2.

6.5 SINGLE CLASS EXTRACTION

Many times, the end user of remote sensing classified output maps is interested in a specific class or a particular land cover only (Foody et al., 2006; Li and Guo, 2010; Li et al., 2011). At the same time, it is not necessary to focus on the other n-number of classes present in the scene. Thus, the analyst has to devote time and effort to extract the information of no use along with the interest class. For an example, if a user is interested to update the transport system, his/her interest will be to extract only road features from the satellite data. Other features like water bodies, agricultural land, and forests will be of no use for that user. To extract the specific land cover from the traditional supervised classification, it is necessary to have the information of all land cover types at the training stage; in other words, classes should be exhaustively defined (Boyd et al., 2006; Foody et al., 2006; Sanchez-Hernandez et al., 2007; Li and Guo, 2010; Li et al., 2011). This not only increases the classification cost and labor, but may also produce substantial error in the output (Foody et al., 2006; Li et al., 2011). Thus, for specific class extraction, conventional supervised classification methods are inappropriate (Foody et al., 2006). Foody et al. (2006) made an attempt to reduce the training set size while extracting the specific class. A very few studies (Kumar et al., 2010; Sengar et al., 2012; Upadhyay et al., 2012, 2013, 2014) have been used to extract specific class from the remote sensing data.

6.5.1 Fuzzy Set Theory Based Classifiers for a Single Class Extraction

The fuzzy set theory based classifiers discussed in Chapter 3 and hybrid classifiers in Chapter 5 have the capability to identify the specific class of interest. In this section, the behavior of fuzzy set theory based classifiers as well as hybrid classifiers for a single information class (or cluster) is discussed. It may be noted that the uniqueness of this approach is that the training data for only that class will be provided. Let us take the membership function of fuzzy c-means (FCM) from Equation (3.11) for extraction of a single class present in the dataset, then $D(x_k, \bar{v}_i) = D(x_k, \bar{v}_j)$ and $\bar{\mu}_{ki} = 1$, which indicates that the membership of all features will be equal to one. Therefore, all the pixels in a digital image will belong to a single class, which is never true. Thus, the FCM algorithm fails for the extraction of a single class from the image.

Further, the possibilistic c-means (PCM) follows the FCM algorithm in the initial iterations. Thus, by considering Equations (3.17) and (3.18), the membership function and bandwidth parameter (η_i) for single information class can be given in Equation (6.7):

$$\bar{\mu}_{ki} = 1 \text{ and } \eta_i = K \times \sum_{k=1}^{N} D(x_k, \bar{v}_i) \Big/ N \tag{6.7}$$

Thus, in the case of PCM, $\bar{\mu}_{ki}$ for single class out of "n" number of classes present in the image can be calculated by from Equation (6.8)

$$\bar{\mu}_{kc} = \frac{1}{1 + \left(D(x_k, \bar{v}_c)/\eta_c\right)^{\frac{1}{(m-1)}}} \tag{6.8}$$

Similarly, the MPCM algorithm also follows the same initial step as mentioned in Equation (6.9):

$$\bar{\mu}_{ki} = 1 \text{ and } \eta_i = \sum_{k=1}^{N} D(x_k, \bar{v}_i) \Big/ N \tag{6.9}$$

whereas the membership for single class can be computed from Equation (6.10):

$$\bar{\mu}_{kc} = \exp\left(\frac{-D(x_k, \bar{v}_c)}{\eta_c}\right) \tag{6.10}$$

By considering the membership function of fuzzy c-means with entropy (FCME) from Equation (5.5), for single class extraction $D(x_k, \bar{v}_i) = D(x_k, \bar{v}_j)$ and $\bar{\mu}_{ki} = 1$. Thus, FCME algorithm is also not suitable for identification of single land cover from the image. Further, for extraction of a single class, the membership values of the noise clustering (NC) classifier can be calculated by substituting $D(x_k, v_i) = D(x_k, v_j) \cong D(x_k, v_c)$ in Equations (3.22) and (3.23). Therefore, for NC classifier the membership values for extraction of single class can be obtained as from Equations (6.11) and (6.12):

$$\bar{\mu}_{kc} = \left[1 + \left(\frac{D(x_k, v_c)}{\delta}\right)^{\frac{1}{m-1}}\right]^{-1} \tag{6.11}$$

and

$$\bar{\mu}_{k,c+1} = \left[\left(\frac{\delta}{D(x_k, v_c)}\right)^{\frac{1}{m-1}} + 1\right]^{-1} \tag{6.12}$$

From Equations (6.11) and (6.12), it is also evident that for a single information class, the membership values for both good cluster and noise cluster remains significant with $\bar{\mu}_{kc} = 1 - \bar{\mu}_{k,c+1}$. Thus, the membership value of a noise point in a good cluster is not forced to one (Dave and Krishnapuram, 1997).

In a similar manner, the membership values and the class center using noise clustering with entropy (NCE) classifier for a single information class can be expressed as from Equations (6.13) and (6.14):

$$\bar{\mu}_{kc} = \frac{\exp\left(-\frac{D(x_k, v_c)}{v}\right)}{\exp\left(-\frac{D(x_k, v_c)}{v}\right) + \exp\left(-\frac{\delta}{v}\right)} \tag{6.13}$$

and

$$\bar{\mu}_{k,c+1} = \frac{\exp\left(-\frac{\delta}{v}\right)}{\exp\left(-\frac{D(x_k, v_c)}{v}\right) + \exp\left(-\frac{\delta}{v}\right)} \tag{6.14}$$

where $\bar{\mu}_{kc} = 1 - \bar{\mu}_{k,c+1}$.

As PCM-S is a PCM derived classifier, it enables the labeling of a pixel to be affected by labels of its neighbors (Singh et al., 2019). Therefore, like PCM classifiers, the PCM-S is also capable for extraction of single class while reducing noisy pixels. The membership function while extracting single class gets modified as mentioned in Equation (6.15).

$$\bar{\mu}_{kc} = \frac{1}{1 + \left[\frac{D(x_k, \bar{v}_c) + \frac{a}{N_R} \sum_{r \in N_k r \neq k} D(x_k, \bar{v}_c)}{\eta_c}\right]^{\frac{1}{m-1}}} \tag{6.15}$$

More details about the PCM-S algorithm can be seen in Chapter 5, Section 5.8.2.

PLICM can also be used for single class extraction; however, its membership function gets modified in comparison PLICM algorithm explained in Chapter 5. The PLICM membership function for extraction of single class appears as mentioned in Equation (6.16).

$$\bar{\mu}_{kc} = \frac{1}{1 + \left[\frac{D(x_k, \bar{v}_c) + G_{kc}}{\eta_c}\right]^{\frac{1}{m-1}}} \tag{6.16}$$

More details about the PLICM algorithm can be seen in Chapter 5, Section 5.8.4.

Further, ADPLICM is also a PCM based classifier. Therefore, the ADPLCM algorithm can also be used for single class extraction. The membership function for single class extraction using ADPLICM appears as mentioned in Equation (6.17).

$$\overline{\mu_{kc}} = \frac{1}{1+\left[\dfrac{D(x_k,\bar{v}_c)+\dfrac{1}{N_R}\sum\limits_{r\in N_k r\neq k}(1-S_{kr})D(x_r,v_c)}{\eta_c}\right]^{\frac{1}{m-1}}} \tag{6.17}$$

More details about the ADPLICM algorithm can be seen in Chapter 5, Section 5.8.6.

MPCM method has the capability to extract single class of interest, but cannot handle complex noise in the image data. To overcome this, modified possibilistic c-means with constraints (MPCM-S) can be used. So, MPCM-S can extract a single class of interest while handling noise present in the image. The membership function of MPCM-S while extracting single class gets modified as given in Equation (6.18). More details about the MPCM-S algorithm can be seen in Chapter 5, Section 5.8.7.

$$\bar{\mu}_{kc} = \exp\left[-\frac{\left\{D(x_k,\bar{v}_c)+\dfrac{a}{N_R}\sum\limits_{r\in N_k r\neq k}D(x_r,v_c)\right\}}{\eta_c}\right] \tag{6.18}$$

As discussed in Chapter 5, the MPLICM method has been introduced to integrate the local spatial and grey level information for preserving the image details as well as to control the effect of neighborhood in PCM. The MPLICM method can also be applied for single class extraction; however, its membership function will get modified as given in Equation (6.19).

$$\bar{\mu}_{kc} = \exp\left[-\frac{\left\{D(x_k,\bar{v}_c)+G_{kc}\right\}}{\eta_c}\right] \tag{6.19}$$

More details about the MPLICM algorithm can be seen in Chapter 5, Section 5.8.8.

The ADMPLICM method has been introduced after incorporation of local similarity measure in MPCM. The ADMPLICM can also be used for single class extraction; however, its membership function gets modified while extracting single class extraction as given in Equation (6.20).

$$\bar{\mu}_{kc} = \exp\left[-\frac{\left\{D(x_k,\bar{v}_c)+\frac{1}{N_R}\sum_{r\in N_k r\neq k}(1-S_{kr})D(x_r,v_c)\right\}}{\eta_c}\right] \tag{6.20}$$

More details about the ADMPLICM algorithm can be seen in Chapter 5, Section 5.8.9.

Thus, it can be concluded that among all the fuzzy set theory based classifiers, only some are capable of single land cover classification. The utility of different fuzzy based classifiers for mixed pixel and for single land cover identification is shown in Table 6.2.

TABLE 6.2
List of Fuzzy Set Theory Based Classifiers for Mixed Pixel and Single Land Cover Identification

Fuzzy Set Theory Based Classifier	Mixed Pixel	Single Class Identification
FCM	✓	–
FCME	✓	–
PCM	✓	✓
PCME	✓	✓
IPCM	✓	–
MPCM	✓	✓
NC	✓	✓
NCWE	✓	✓
PCM-S	✓	✓
FLICM	✓	–
PLICM	✓	✓
ADFLICM	✓	–
ADPLICM	✓	✓

6.6 CONCEPT FOR MONO-/BI-SENSOR REMOTE SENSING DATA PROCESSING

Generally extreme homogeneous land cover like water bodies (e.g., reservoirs) can easily be identified uniquely by the aforementioned classifiers from single date imagery. However, other features and single land cover are less homogeneous in nature; crops or vegetation classes having similar spectral similarity with other classes at different times can be identified using temporal images. This is due to significant changes in spectral response of such class pixels in contrast to classes of no variation or high correlation, thereby making them highly identifiable as a single land cover through temporal images. As a first step, this requirement of temporal data can be attempted from mono-sensors. In case of non-availability of the required temporal data dates, the bi-sensor approach can be used to meet the temporal gaps. These can be filled whether using optical remote sensing or microwave remote sensing or both of them together.

6.7 SUMMARY

From this chapter it can be inferred that the temporal analysis of dynamic land cover classes is capable to identify them uniquely from the rest. The inclusion of spectral indices as well as capable fuzzy classifiers has made it possible to extract single land cover from a remote sensing image.

BIBLIOGRAPHY

Alcantara, C., Kuemmerle, T., Prishchepov, A.V. and Radeloff, V.C., 2012. Mapping abandoned agriculture with multi-temporal MODIS satellite data. *Remote Sensing of Environment*, 124, 334–347.

Atzberger, C. and Eilers, P.H.C., 2011. A time series for monitoring vegetation activity and phenology at 10-daily time steps covering large parts of South America. *International Journal of Digital Earth*, 4 (5), 365–386.

Birth, G.S. and McVey, G., 1968. Measuring the color of growing turf with a reflectance spectroradiometer. *Agronomy Journal*, 60, 640–643.

Blaes, X., Vanhalle, L. and Defourny, P., 2005. Efficiency of crop identification based on optical and SAR image time series. *Remote Sensing of Environment*, 96, 352–365.

Boyd, D.S., Sanchez-Hernandez, C. and Foody, G.M., 2006. Mapping a specific class for priority habitats monitoring from satellite sensor data. *International Journal of Remote Sensing*, 27 (13), 2631–2644.

Bruzzone, L. and Serpico, S.B., 2000. A technique for feature selection in multiclass problems. *International Journal of Remote Sensing*, 21 (3), 549–563.

Campbell, J.B., 1987. *Introduction to Remote Sensing*. New York: The Guilford Press.

Dadhwal, V.K., Singh, R.P., Dutta, S. and Parihar, J.S., 2002. Remote sensing based crop inventory: A review of Indian experience. *Tropical Ecology*, 43 (1), 107–122.

Dave, R.N. and Krishnapuram, R., 1997. Robust clustering methods: Unified view. *IEEE Transactions on Fuzzy Systems*, 5, 270–293.

Deering, D.W., Rouse, J.W., Haas, R.H. and Schell, J.A., 1975. Measuring "Forage Production" of grazing units from Landsat MSS data. In: *Proceedings of the 10th International Symposium on Remote Sensing of Environment*, Environment Research Institute of Michigan, Ann Arbor, Michigan, vol. II, pp. 1169–1178.

Foody, G.M., Mathur, A., Sanchez-Hernandez, C. and Boyed, D.S., 2006. Training set size requirements for the classification of a specific class. *Remote Sensing of Environment*, 104, 1–14.

GAO, B.C., 1996. NDWI—A normalized difference water index for remote sensing of vegetation liquid water from space. *Remote Sensing of Environment*, 58, 257–266.

Ghosh, S.K., 2013. *Digital Image Processing*. India: Narosa Publishing House Pvt. Ltd.

Goetz, S.J., Fiske, G.J. and Bunn, A.G., 2006. Using satellite time-series data sets to analyze fire disturbance and forest recovery across Canada. *Remote Sensing of Environment*, 101, 352–365.

Gonçalves, R.V., Jr, Zullo, J., Romani, L.A.S., Nascimento, C.R. and Traina, A.J.M., 2012. Analysis of NDVI time series using cross-correlation and forecasting methods for monitoring sugarcane fields in Brazil. *International Journal of Remote Sensing*, 33 (15), 4653–4672.

Hall, D.K., Riggs, G.A. and Solomonson, V.V., 1995. Development of methods for mapping global snow cover using moderate resolution imaging spectroradiometer data. *Remote Sensing of Environment*, 54, 127–140.

Hilker, T., Wulder, M.A., Coops, N.C., Linke, J., McDermid, G., Masek, J.G., Gao, F. and White, J.C., 2009. A new data fusion model for high spatial- and temporal-resolution mapping of forest disturbance based on Landsat and MODIS. *Remote Sensing of Environment*, 113, 1613–1627.

Huete, A.R. and Liu, H.Q., 1994. An error and sensitivity analysis of the atmospheric and soil-correcting variants of the NDVI for the MODIS-EOS. *IEEE Transactions on Geoscience and Remote Sensing*, 32, 897–905.

Huete, A.R., Liu, H.Q., Batchily, K. and Leeuwenvan, W., 1997. A comparison of vegetation indices over a global set of TM images for EOS-MODIS. *Remote Sensing of Environment*, 59, 440–451.

Jensen, J.R., 1996. *Introductory Digital Image Processing: A Remote Sensing Perspective*. 2nd Edition. Upper Saddle River, NJ: Prentice Hall.

Jiang, Z., Huete, A.R., Didan, K. and Miura, T., 2008. Development of a two-band enhanced vegetation index without a blue band. *Remote Sensing of Environment*, 112, 3833–3845.

Kaufman, Y.J. and Tanre, D., 1992. Atmospherically resistance vegetation index (ARVI) for EOS-MODIS. *IEEE Transactions on Geoscience and Remote Sensing*, 30, 261–270.

Key, C.H. and Benson, N.C., 2006. Landscape assessment: Ground measure of severity, the composite burn index; and remote sensing of severity, the Normalized Burn Ratio. In Lutes, D.C., Keane, R.E., Caratti, J.F., Key, C.H., Benson, N.C., Sutherland, S., and Gangi, L.J. (eds.). *FIREMON: Fire Effects Monitoring and Inventory System*. USDA Forest Service, Rocky Mountain Research Station, Ogden, UT. Gen. Tech. Rep. RMRS-GTR-164-CD: LA1-51.

Kumar, A., Ghosh, S.K. and Dadhwal, V.K., 2010. ALCM: Automatic land cover mapping. *Journal of Indian Society of Remote Sensing*, 8 (2), 239–245. http://dx.doi.org/10.1007/s12524-010-0030-x.

Li, W. and Guo, Q., 2010. A maximum entropy approach to one class classification of remote sensing imagery. *International Journal of Remote Sensing*, 31 (8), 2227–2235.

Li, W., Guo, Q. and Elkan, C., 2011. A positive and unlabeled learning algorithm for one-class classification of remote-sensing data. *IEEE Transactions on Geoscience and Remote Sensing*, 49 (2), 717–725.

Maselli, F., 2011. Use of MODIS NDVI data to improve forest-area estimation. *International Journal of Remote Sensing*, 32 (11), 6379–6393.

Masialeti, I., Egbert, S., and Wardlow, B.D., 2010. A comparative analysis of phenological curves for major crops in Kansas. *GIScience & Remote Sensing*, 47 (2), 241–259. https://doi.org/10.2747/1548-1603.47.2.241.

Millward, A.A., Piwowar, J.M. and Howarth, P.J., 2006. Time-series analysis of medium-resolution, multisensor satellite data for identifying landscape change. *Photogrammetric Engineering & Remote Sensing*, 72 (6), 653–663.

Morton, D.C., DeFries, R.S., Nagol, J., Souza, C.M., Kasischke, E.S., Hurtt, G.C. and Dubayah, R., 2011. Mapping canopy damage from understory fires in Amazon forests using annual time series of Landsat and MODIS data. *Remote Sensing of Environment*, 115, 1706–1720.

Pan, Y., Li, L., Zhang, J., Liang, S., Zhu, X. and Sulla, M.D., 2012. Winter wheat area estimation from MODIS-EVI time series data using the Crop Proportion Phenology Index. *Remote Sensing of Environment*, 119, 232–242.

Pennec, A., Gond, V. and Sabatier, D., 2011. Tropical forest phenology in French Guiana from MODIS time series. *Remote Sensing Letters*, 2 (4), 337–345.

Potgieter, A.B., Apan, A., Hammer, G. and Dunn, P., 2010. Early-season crop area estimates for winter crops in NE Australia using MODIS satellite imagery. *ISPRS Journal of Photogrammetry and Remote Sensing*, 65, 380–387.

Powell, S.L., Cohen, W.B., Sean P. Healey, S.P., Kennedy, R.E., Gretchen G. Moisen, G.G., Pierce, K.B. and Ohmann, J.L., 2010. Quantification of live aboveground forest biomass dynamics with Landsat time-series and field inventory data: A comparison of empirical modeling approaches. *Remote Sensing of Environment*, 114, 1053–1068.

Pringle, M.J., Denham, R.J. and Devadas, R., 2012. Identification of cropping activity in central and southern Queensland, Australia, with the aid of MODIS MOD13Q1 imagery. *International Journal of Applied Earth Observation and Geoinformation*, 19, 276–285.

Rouse, J.W., Haas, R.H., Schell, J.A. and Deering, D.W., 1973. Monitoring vegetation systems in the Great Plains with ERTS. *Third ERTS Symposium*, NASA SP-351, Washington DC, December 10–14, 1973, 309–317.

Running, S.W., Justice, C.O., Salomonson, V., Hall, D., Barker, J., Kaufmann, Y.J., Strahler, A.H. et al., 1994. Terrestrial remote sensing science and algorithms planned for EOS/MODIS. *International Journal of Remote Sensing*, 5, 3587–3620.

Sakamoto, T., Yokozawa, M., Toritani, H., Shibayama, M., Ishitsuka, N. and Oho, H., 2005. A crop phenology detection method using time-series MODIS data. *Remote Sensing of Environment*, 96, 366–374.

Sakamoto, T., Nguyen, N.V., Kotera, A., Ohno, H., Ishitsuka, N. and Yokozawa, M., 2007. Detecting temporal changes in the extent of annual flooding within the Cambodia and the Vietnamese Mekong Delta from MODIS time-series imagery. *Remote Sensing of Environment*, 109, 295–313.

Sanchez-Hernandez, C., Boyed, D.S. and Foody, G.M., 2007. One-class classification for mapping a specific land-cover class: SVDD classification of Fenland. *IEEE Transactions on Geoscience and Remote Sensing*, 45 (4), 1061–1073.

Schubert, P., Lagergren, F., Aurela, M., Christensen, T., Grelle, A., Heliasz, M., Klemedtsson, L. et al., 2012. Modeling GPP in the Nordic forest landscape with MODIS time series data—Comparison with the MODIS GPP product. *Remote Sensing of Environment*, 126, 136–147.

Sengar S.S., Kumar, A., Ghosh, S.K., Wason, H.R. and Roy, P.S., 2012. Liquefaction identification using class-based sensor independent approach based on single pixel classification after 2001 Bhuj, India earthquake. *Journal of Applied Remote Sensing*, 6 (1), 063531. https://doi.org/10.1117/1.JRS.6.063531.

Singh, A. and Kumar, A., 2019. Fuzzy based approach to incorporate spatial constraints in possibilistic *c*-means algorithm for remotely sensed imagery. In: *International Conference on Sustainable Computing in Science, Technology & Management*, Jaipur, February 20, 2019. https://dx.doi.org/10.2139/ssrn.3354465.

Steven, M.D., Malthus, T.J., Baret, F., Xu, H. and Chopping, M.J., 2003. Intercalibration of vegetation indices from different sensor systems. *Remote Sensing of Environment*, 88 (4), 412–422. https://doi.org/10.1016/j.rse.2003.08.010.

Swain, P.H. and Davis, S.M., (eds.), 1978. *Remote Sensing: The Quantitative Approach* New York: McGraw-Hill.

Tingting, L. and Chuang, L., 2010. Study on extraction of crop information using time-series MODIS data in the Chao Phraya Basin of Thailand. *Advances in Space Research*, 45 (6), 775–784.

Tso, B. and Mather, P.M., 2009. *Classification Methods for Remotely Sensed Data*. Boca Raton, FL: CRC Press, Taylor & Francis Group.

Upadhyay, P., Kumar, A., Roy, P.S., Ghosh, S.K. and Gilbert, I., 2012. Effect on specific crop mapping using WorldView-2 multispectral add-on bands: Soft classification approach. *Journal of Applied Remote Sensing*, 6 (1), 063524. https://doi.org/10.1117/1.JRS.6.063524.

Upadhyay, P., Ghosh, S.K. and Kumar, A., 2013. Moist deciduous forest identification using temporal MODIS data: A comparative study using fuzzy based classifiers. *Ecological Informatics* (*Elsevier*), 18, 117–130, http://dx.doi.org/10.1016/j.ecoinf.2013.07.002.

Upadhyay, P., Ghosh, S. K., Kumar, A., Krishna Murthy, Y.V.N. and Raju, P.L.N., 2014. Moist deciduous forest identification using MODIS temporal indices data. *International Journal of Remote Sensing*, 35 (9), 3177–3196.

Upadhyay, P., Ghosh, S.K. and Kumar, A., 2016. Temporal MODIS data for identification of wheat crop using noise clustering soft classification approach. *Geocarto International*, 31 (3), 278–295, http://dx.doi.org/10.1080/10106049.2015.1047415.

Wang, Q. and Tenhunen, J.D., 2004. Vegetation mapping with multi temporal NDVI in North Eastern China Transect (NECT). *International Journal of Applied Earth Observation and Geoinformation*, 6, 17–31.

Wardlow, B.D. and Egbert, S.L., 2008. Large-area crop mapping using time-series MODIS 250 m NDVI data: An assessment for the U.S. Central Great Plains. *Remote Sensing of Environment*, 112, 1096–1116.

Wardlow, B.D. and Egbert, S.L., 2010. A comparison of MODIS 250-m EVI and NDVI data for crop mapping: A case study for southwest Kansas. *International Journal of Remote Sensing*, 31 (3), 805–830.

Wardlow, B.D., Egbert, S.L. and Kastens, J.H., 2007. Analysis of time-series MODIS 250 m vegetation index data for crop classification in the U.S. Central Great Plains. *Remote Sensing of Environment*, 108, 290–310.

Xiao, X., Boles, S., Frolking, S., Li, C., Babu, J.Y., Salas, W. and Moore, B., 2006. Mapping paddy rice agriculture in South and Southeast Asia using multi-temporal MODIS images. *Remote Sensing of Environment*, 100, 95–113.

7 Assessment of Accuracy for Soft Classification

7.1 INTRODUCTION

Remote sensing is one of the most important tools for generating land use/land cover information as thematic maps of the earth's surface. But it is always important to evaluate thematic maps generated from remote sensing data as valuable information using assessment of accuracy methods (Foody, 2002). From the past few decades, remote sensing data is being used for land use/land cover information from regional to global scales at the sub-pixel level (Fisher and Pathirana, 1990; Cross et al., 1991; Gutman and Ignatov, 1998; Carpenter et al., 1999). The quality assessment of sub-pixel classification outputs is difficult using the methods used for hard classified outputs. Thematic maps prepared from any of the classifiers are not complete without performing the assessment of accuracy (Tortora, 1978; Jensen, 1996; Congalton and Green, 1999).

Hard classification provides labels a pixel to one and only one class. A typical strategy for assessing the accuracy of hard classification is to follow a statistically sound sampling design to select a sample of testing pixels. It is thus used to determine whether the class assigned to that pixel matches the actual class represented by pixels on the reference data or not. The sample data is used to generate an error matrix, which is then used to derive various accuracy measures (Congalton, 1991; Congalton and Green, 1999).

In soft classification, mixed pixels are processed to extract classes present in that pixel. The mixed pixels are often dominant in remotely sensed data of heterogeneous land cover (Liu and Wu, 2005; Xu et al., 2010). The outputs generated for each class using the soft classification methods are known as fraction outputs. These fractional outputs hold membership values. Since these membership values are float values between 0 and 1, the conventional error matrix method falls short in using these membership values for assessing the accuracy of soft classification (Pontius and Cheuk, 2006).

As described earlier, the evaluation of hard classified outputs can be conducted through assessment of accuracy using confusion matrix and its derived measures (Congalton, 1991; Stehman and Czaplewski, 1998; Congalton and Green, 1999). Confusion matrix takes each sample input from classified and reference as an integer value and hence is appropriate only for hard classifications. Fraction images generated through soft classifications hold membership values of each class at a per pixel level. For assessment of accuracy of fraction images, there is no standard procedure established yet. So, there is high demand for sub-pixel assessment of classified fraction outputs generated from remote sensing studies (Latifovic and Olthof, 2004; Shabanov et al., 2005; Okeke and Karnieli, 2006; Ozdogan and Woodcock, 2006).

For assessment of accuracy of soft classifications, various methods have been suggested (Gopal and Woodcock, 1994; Foody, 1996; Binaghi et al., 1999; Congalton and Green, 1999; Townsend, 2000; Lewis and Brown, 2002; Pontius Jr. and Cheuk, 2006). One of the methods similar to confusion matrix is the fuzzy error matrix (Binaghi et al., 1999). This is one of the most commonly used approaches and is based on fuzzy set theory and represents a generalization of traditional confusion matrix. Even though fuzzy set theory has a sound theoretical basis, it has not been generally adopted as a standard accuracy method for soft classifications. Some of the reasons for this have been highlighted as counter-intuitive characteristics (Pontius Jr. and Cheuk, 2006).

Mainly, the comparison of soft evaluation methods with the conventional confusion matrix should be consistent, and its total of membership grades should match the marginal totals. Secondly, there should be interpretable information on the assessment of soft classification outputs. A composite operator for generating cross-comparison matrix that displays some of the aforesaid necessary features can be considered as a method for comparison of sub-pixel maps (Pontius Jr. and Cheuk, 2006; Pontius Jr. and Connors, 2006). Nevertheless, single or composite operators generate elements of fuzzy error matrix and use of off-diagonal cells, and with their interpretation shown clearly.

For conducting assessment of accuracy of a soft classified output, a different type of error matrix known as fuzzy error matrix (FERM) has been proposed by the Binaghi et al. (1999). A further advanced form of FERM was proposed by Silvan and Wang (2008), known as the sub-pixel confusion-uncertainty matrix (SCM). The fuzzy error matrix can be used to generate the accuracy of soft classifier outputs, while these outputs are in the form of fraction images. FERM is similar to conventional error matrix; however, it uses fractional images and fractional values (real numbers) as input to measure accuracy, while the conventional method uses integer values.

The correspondence between classified and reference datasets in FERM is calculated while applying single operators like *minimum* (MIN). But other operators can also be applied like *least* (LEAST) and *product* (PROD). There are methods to generate composite operators like MIN-MIN, MIN-LEAST, and MIN-PROD in modified FERM matrix (Pontius and Cheuk, 2006; Silván and Wang, 2008).

The limitation of FERM and its modified versions are due to the region that the classified and reference data should be in form of fraction images. Many times, it is not possible to have reference data as fraction images at finer spatial resolution. In those cases, some of the researchers proposed absolute methods to have quantitative assessment of fraction classified outputs. Dehghan and Ghassemian (2006) proposed entropy, which measures uncertainty from the samples of classified output as a measure to assess accuracy. Since in generation of entropy, sample membership of classified output only is required; this entropy is also called an absolute measure of uncertainty. It is absolute as it does not refer any reference data to measure the uncertainty. Since from entropy parameters, direct assessment of accuracy cannot be represented, it is called an indirect method of assessment of accuracy. Thus, the entropy method as an absolute method of accuracy measure can be used even if no reference data is available for assessing the accuracy. The interpretation of entropy includes ideas such as higher entropy implies higher uncertainty and vice versa. Kumar and Dadhwal (2010) also have proven the advantage of entropy measures for assessment of accuracy.

7.2 GENERATION OF TESTING DATA

Assessment of accuracy is dependent on the quality of the reference sample data while comparing classified image samples. A classified image is the output product generated applying a classification algorithm. As discussed in Chapter 2, reference data may be collected from different sources like existing maps, field survey data, or classified output from finer spatial resolution images. There can be combinations of these sources for generating a reference sample dataset. Assessment of accuracy of fraction images as soft classified output can be conducted using soft reference data sets. Generation of soft reference data may be impossible from ground observations. In this chapter, the image-to-image based assessment of accuracy method has been explained for assessing the soft classified outputs. In order to assess the accuracy of soft classified output, there is a requirement of soft reference data. The easiest way to generate soft reference data is to use fine resolution images of the same area acquired at same time as of coarse resolution classified images. From this fine resolution image, membership value of each class from each pixel has been generated as reference data. Fine resolution image will be covering more pixels as reference sample pixels within one coarser pixel of classified image. So membership value of one pixel at output pixel will be assessed by membership values of more than one pixel from reference image.

While performing the image-to-image registration, there is "one to one" correspondence of pixels between classified and reference images. Therefore, pixels from classified and reference images represent the same area on the ground and thus have adjunct overlap. Therefore, after registering the two images corresponding to one classified pixel from classified image, there will be more than one pixel from the reference image. On these fraction images from classified and reference images, various operators can be applied to generate various types of fuzzy error matrixes, which are discussed in coming sections.

7.3 METHODS FOR ASSESSMENT OF ACCURACY OF SOFT CLASSIFIED OUTPUTS

Assessment of accuracy of classified output is necessary to obtain the quality of results. For soft classified output, modified accuracy assessment methods have been suggested. As mentioned in the previous section, an image-to-image based assessment of accuracy method is used for accuracy of classification of a coarser resolution image using the classified outputs of a finer resolution image of similar time and geographical area. In this section, various accuracy assessment techniques for soft classified outputs are described. Using different operators, various image-to-image based methods for assessment of soft classification such as fuzzy error matrix, composite operators, and sub-pixel confusion matrix are described. Further, some other methods as absolute independent assessment methods are entropy, variance–covariance, correlation, root mean square error, and receiver operating characteristics (ROC) methods are also described.

7.3.1 Fuzzy Error Matrix and Other Associated Operators

Since there is no standard procedure for assessing soft classified outputs, sometimes the outputs are defuzzified to produce hard outputs. However, due to defuzzification, loss of information occurs and the purpose of classifying the mixed pixels is nullified. Therefore, to preserve the result of soft classified outputs, image-to-image based accuracy with various operators has been proposed. In this section, fuzzy error matrix and its variants are discussed in detail.

7.3.1.1 Fuzzy Error Matrix

To handle mixed pixels, soft classification techniques have to apply, as remote sensing derived land cover maps and associated statistics are useful. So, there should be a method for quantitatively assessing accuracy of soft classified output (Meyer and Werth, 1990). In the 1970s–1980s, classification accuracy assessment was an afterthought rather than an integral part of many remote sensing studies. In a number of studies, the classification accuracy report has been expressed simply as a single number (e.g., 86%). These overall accuracy assessments do not provide locational accuracy. In another way, only the total assessment of all classes is considered without its location. Accuracy assessment that is not site-specific yields very high accuracy, but gives misleading results when all the errors are located in a region.

In accuracy assessment of classification results, two sources of data are compared: (1) classification map derived from remote-sensing data and (2) reference test date. An error matrix provides the relationship between classified and reference data sets. An error matrix is a square array in size, which is laid out in rows and columns. This matrix expresses the number of samples units (i.e., pixels, clusters of pixels, or polygons) assigned to a particular category relative to the actual category as verified in the field. Reference data is represented by columns while classified data is represented by rows from the remotely sensed data. By an error matrix, accuracy of each category is clearly described, which is a very effective way to represent accuracy. As from error matrix along with overall accuracy, both the errors of inclusion (commission errors) and errors of exclusion (omission errors) can also be calculated.

For the assessment of accuracy of soft classified data, FERM, which is based on the fuzzy set theory, is one of the most basic approaches (Binaghi et al., 1999). The layout of FERM is similar to the traditional error matrix that is used for assessing the accuracy of hard classification. The exception is that the elements of the FERM can be any non-negative real numbers instead of non-negative integer numbers. The row of the FERM generally defines the soft classified data and the column defines the soft reference data. The elements of FERM represent the class proportion corresponding to reference data (i.e., soft reference data) and classified outputs (i.e., soft classified image), respectively. The fuzzy minimum operator (MIN) is used to construct the FERM and to determine the matrix elements in which the degree of membership in fuzzy interaction in between the classified and reference partition $(C_i \cap R_j)$ is computed as (Binaghi et al., 1999; Stehman et al., 1998) in Equation (7.1):

$$M(i,j) = |C_i \cap R_j| = \sum_{x \in X} \min(s_{ki}, r_{kj}) \tag{7.1}$$

where R_j and C_i are the sets of reference and classification data assigned to classes j and i, X is the testing sample data set, x is a testing sample in X, and s_{ki} and r_{kj} denote the classified and reference grades of class i and j respectively at pixel k and $i, j = 1, 2, \dots c$.

The layout of the FERM is shown in Table 7.1.

The derived indices from the FERM are the overall, producer, and user accuracy measures. The overall accuracy (OA) can be calculated by Equation (7.2):

$$OA = \frac{\sum_{i=1}^{c} M_{(i,i)}}{\sum_{i=1}^{c} R_i} = P_O \tag{7.2}$$

where P_O is the observed proportion of agreement.

TABLE 7.1
Layout for a Fuzzy Error Matrix

	Soft Reference Data				
Soft Classification	**Class 1**	**Class 2**	**...**	**Class c**	**Total Grades**
Class 1	M(1,1)	M(1,2)	...	M(1,c)	C1
Class 2	M(2,1)	M(2,2)	...	M(2,c)	C2
...	.	.	.	.	.
	.	.	.	.	.
	.	.	.	.	.
Class c	M(c,1)	M(c,2)	...	M(c,c)	Cc
Total Grades	R1	R2	...	Rc	

Note: Definition of terms: $M_{(i,j)}$ is the member of FERM in the i-th class in soft classified output and jth class of soft reference data, C_i is the sum of class proportions of class i in the classified output, and R_j is the sum of class proportions of class j from the reference data.

The user's (UA_i) and producer's (PA_i) accuracies of class i may be computed as Equation (7.3),

$$UA_i = \frac{M_{(i,i)}}{C_i} \text{ and } PA_i = \frac{M_{(i,i)}}{R_i} \tag{7.3}$$

while the average user's (AA_u) and average producer's (AA_p) accuracies may be computed from Equation (7.4),

$$AA_u = \frac{\sum_{i=1}^{c} UA_i}{c} \text{ and } AA_p = \frac{\sum_{i=1}^{c} PA_i}{c} \tag{7.4}$$

and expected proportion of agreement (P_E) can be expressed as in Equation (7.5),

$$P_E = \frac{\sum_{i=1}^{c} R_i C_i}{\left(\sum_{i=1}^{c} R_i\right)^2} \tag{7.5}$$

and kappa coefficient (κ) of agreement as in Equation (7.6),

$$\kappa = \frac{(P_O - P_E)}{1 - P_E} \tag{7.6}$$

FERM accounts for the diagonalization characteristic (a perfect matching case, agreement up to pixel level) of a sub-pixel confusion matrix. The derived indices of FERM (producer, user, and overall accuracy) are based on diagonal elements and total grades of reference and assessed data sets (Silván-Cárdenas and Wang, 2008) and do not account for off-diagonal elements, which represent disagreement.

7.3.1.2 Composite Operator Based FERM

Although FERM is the most basic approach for assessment of accuracy of soft classification, it is not used as a standard measure of accuracy. The reason is that the cross-comparison in FERM is not consistent with the traditional confusion matrix. For the cross-comparison to be consistent, it must have a diagonal matrix when a map is compared to itself, and its marginal totals must match with the total of membership grades from the reference and assessed data. However, the composite operators based accuracy measure proposed by Pontius and Cheuk (2006) can be used for the computation of cross-comparison matrix. For these composite operators, certain fundamental properties on agreement and disagreement have been established so that meaningful matrix entries can be made (Silván-Cárdenas and Wang, 2008).

The agreement and disagreement measures for the composite operator are given in Equations (7.7) to (7.9):

$$C\left(s_{ki}, r_{kj}\right) = \begin{cases} A\left(s_{ki}, r_{kj}\right), \text{if } i = j \\ D\left(s'_{ki}, r'_{kj}\right), \text{if } i \neq j \end{cases} \tag{7.7}$$

$$s'_{ki} = s_{ki} - \min\left(s_{ki}, r_{ki}\right) \tag{7.8}$$

$$r'_{kj} = r_{kj} - \min\left(s_{kj}, r_{kj}\right) \tag{7.9}$$

where A and D denote the agreement and disagreement operators, respectively, s_{ki} and r_{kj} denote the classified and reference grades of class i and j, respectively, at pixel k, and s'_{ki} and r'_{kj} denote the over and underestimation errors at pixel k.

The operator A and D satisfies the basic properties as mentioned in Table 7.2 for agreement and disagreement measures.

The MIN-MIN, MIN-PROD, MIN-LEAST composite operators (Pontius and Cheuk, 2006; Silván-Cárdenas and Wang, 2008) are used for the assessment of soft classified outputs. These composite operators are derived from three basic operators, viz. minimum operator (MIN), product operator (PROD), and LEAST operator. The MIN operator is a fuzzy set intersection operator, and it measures the maximum sub-pixel class overlap, while the PROD operator measures the expected sub-pixel class overlap, and the LEAST operator measures minimum possible sub-pixel class overlap between classified and referenced sub-pixel partition. Expressions for basic and composite operators are listed in Tables 7.3 and 7.4, respectively.

Since the MIN operator satisfies all basic properties mentioned in Table 7.2, all composite operators use MIN operator for an agreement case. The MIN-PROD operator uses MIN for diagonal (agreement) and a normalized PROD for off-diagonal (disagreement) cells, thus combining the fuzzy set view with a probabilistic view (Silván-Cárdenas and Wang, 2008). The MIN-MIN operator uses MIN for both

TABLE 7.2
Basic Properties for Agreement and Disagreement Measures

Property	Definition	Agreement	Disagreement
Commutativity	$C(s,r) = C(r,s)$	Yes	Yes
Positivity	$s > 0 \vee r > 0 \Rightarrow C(s,r) > 0$	Yes	Yes
Nullity	$s = 0 \wedge r = 0 \Rightarrow C(s,r) = 0$	Yes	Yes
Upper Bound	$C(s,r) \leq C(r,r)$	Yes	No
Homogeneity	$C(as,ar) = aC(r,s)$	Yes	Yes

Source: Reprinted from *Remote Sens. Environ.*, 112, Silván-Cárdenas, J.L. and Wang, L., 1081–1095, Copyright 2008 [2020], with permission from Elsevier.

Note: $C(s,r)$ denotes a comparison (agreement and disagreement) measure between grades s and r, and a is a positive number.

TABLE 7.3
Three Basic Operators

Operator ID	Form	Traditional Interpretation	Sub-pixel Interpretation
MIN	$\min(s_{ki}, r_{kj})$	Fuzzy Set Intersection	Maximum Overlap
PROD	$s_{ki} \times r_{kj}$	Joint Probability	Expected Overlap
LEAST	$\max(s_{ki} + r_{kj} - 1, 0)$	Minimum Overlap	Minimum Overlap

Source: Reprinted from *Remote Sens. Environ.*, 112, Silván-Cárdenas, J.L. and Wang, L., Sub-pixel confusion–uncertainty matrix for assessing soft classifications, 1081–1095, Copyright 2008 [2020], with permission from Elsevier.

Note: s_{ki} and r_{kj} denote the classified and reference grades of class i and j, respectively, at pixel k.

TABLE 7.4
Three Composite Operators

Operator ID	Agreement $(i = j)$	Disagreement $(i \neq j)$	Sub-Pixel Confusion
MIN-MIN	$\min(s_{ki}, r_{ki})$	$\min(s'_{ki}, r'_{kj})$	Constrained maximum
MIN-PROD	$\min(s_{ki}, r_{ki})$	$\dfrac{s'_{ki} \times r'_{kj}}{\sum_{i=1}^{c} r'_{ni}}$	Constrained expected
MIN-LEAST	$\min(s_{ki}, r_{ki})$	$\max\left(s'_{ki} + r'_{kj} - \sum_{i=1}^{c} r'_{ki}, 0\right)$	Constrained minimum

Source: Reprinted from *Remote Sens. Environ.*, 112, Silván-Cárdenas, J.L. and Wang, L., Sub-pixel confusion–uncertainty matrix for assessing soft classifications, 1081–1095, Copyright 2008 [2020], with permission from Elsevier.

Note: s_{ki} and r_{ki} denote the classified and reference grades of class i at pixel k.
s'_{ki} *and* r'_{ni} *denote the over- and underestimation errors of class i at pixel k.*

agreement and disagreement case. Similarly, the MIN-LEAST operator uses MIN for diagonal cells and normalized LEAST for off-diagonal cells. The solution corresponds to different types of sub-pixel class overlap by the aforementioned composite operators constrained to unmatched sub-pixel fraction.

7.3.1.3 Sub-Pixel Confusion-Uncertainty Matrix (SCM)

Although it has been observed that the MIN operator is an appropriate candidate for the measure of agreement for sub-pixel confusion matrix, it fails when one accounts for the measure of disagreement. This can be solved by using composite operators based measures. However, this disagreement between off-diagonal elements produces uncertainty in sub-pixel distribution, leading to an underspecified problem termed as the sub-pixel area allocation problem. To account for this problem, Silván-Cárdenas and Wang (2008) proposed a cross-comparison matrix known as

TABLE 7.5
General Structure of SCM

Soft Classification	Soft Reference Data Class 1	Class 2	...	Class c	Total Grades
Class 1	P_{11}	$P_{12} \pm U_{12}$	...	$P_{1c} \pm U_{1c}$	$P_{1+} \pm U_{1+}$
Class 2	$P_{21} \pm U_{21}$	P_{22}	...	$P_{2c} \pm U_{2c}$	$P_{2+} \pm U_{2+}$
...	.	.	.	.	.
	.	.	.	.	.
	.	.	.	.	.
Class c	$P_{c1} \pm U_{c1}$	$P_{c2} \pm U_{c2}$	...	P_{cc}	$P_{c+} \pm U_{c+}$
Total Grades	$P_{+1} \pm U_{+1}$	$P_{+2} \pm U_{+2}$	...	$P_{+c} \pm U_{+c}$	$P_{++} \pm U_{++}$

Source: Reprinted from *Remote Sens. Environ.*, 112, Silván-Cárdenas, J.L. and Wang, L., Sub-pixel confusion–uncertainty matrix for assessing soft classifications, 1081–1095, Copyright 2008 [2020], with permission from Elsevier.

the sub-pixel confusion-uncertainty matrix (SCM). It uses the confusion intervals in terms of center value ± maximum error to account for this uncertainty. These confusion intervals express the possible confusion among classes and are formed by the MIN-MIN and MIN-LEAST composite operators. For the unique solution of an area allocation problem, these confusion intervals should be tight.

Silván-Cárdenas and Wang (2008) represented the confusion interval in the form $P_{ij} \pm U_{ij}$, where P_{ij} represents the center value of the interval and U_{ij} interval half-width (Table 7.5). These values are computed as Equations 7.10 and 7.11 respectively:

$$P_{ij} = \frac{P_{ij}^{\text{MIN-MIN}} + P_{ij}^{\text{MIN-LEAST}}}{2} \tag{7.10}$$

$$U_{ij} = \frac{P_{ij}^{\text{MIN-MIN}} - P_{ij}^{\text{MIN-LEAST}}}{2} \quad \text{where } i, j = 1, 2, \ldots c \tag{7.11}$$

where P_{ij} is the overall agreement-disagreement measure between classified class i and reference class j, U_{ij} is the uncertainty measure between classified class i and reference class j, P_{i+} is the marginal row sum of P_{ij}, U_{i+} is the marginal row sum of U_{ij} for class i, P_{+i} is the marginal column sum of P_{ij}, U_{+i} is the marginal column sum of P_{ij} for class i, and P_{++} and U_{++} are the total sum of P_{ii} and U_{ij}, respectively.

The accuracy indices derived from the SCM have been given in Equations (7.12) to (7.21):

Overall accuracy (OA):

For center value,

$$OA = \frac{P_{++} \sum_{i=1}^{c} P_{ii}}{P_{++}^2 - U_{++}^2} = P_O \tag{7.12}$$

and

for uncertainty,

$$OA = \frac{U_{++}\sum_{i=1}^{c} P_{ii}}{P_{++}^2 - U_{++}^2} = U_O \tag{7.13}$$

where $P_O \pm U_O$ is the observed proportion of agreement.

For i-th user accuracy (UA_i):

For center value,

$$UA_i = \frac{P_{ii}P_{i+}}{P_{i+}^2 - U_{i+}^2} \tag{7.14}$$

and

for uncertainty,

$$UA_i = \frac{P_{ii}U_{i+}}{P_{i+}^2 - U_{i+}^2} \tag{7.15}$$

For i-th producer accuracy (PA_i):

For center value,

$$PA_i = \frac{P_{ii}P_{+i}}{P_{+i}^2 - U_{+i}^2} \tag{7.16}$$

and

for uncertainty,

$$PA_i = \frac{P_{ii}U_{+i}}{P_{+i}^2 - U_{+i}^2} \tag{7.17}$$

Expected proportion of agreement (P_E):

For center value,

$$P_E = \sum_{i=1}^{c} \frac{\left(P_{++}^2 + U_{++}^2\right)\left(P_{+i}P_{i+} + U_{+i}U_{i+}\right) - 2P_{++}U_{++}\left(U_{+i}P_{i+} + P_{+i}U_{i+}\right)}{\left(P_{++}^2 - U_{++}^2\right)^2} \tag{7.18}$$

and

for uncertainty (U_E),

$$U_E = \sum_{i=1}^{c} \frac{2P_{++}U_{++}\left(P_{+i}P_{i+} + U_{+i}U_{i+}\right) - \left(P_{++}^2 + U_{++}^2\right)\left(U_{+i}P_{i+} + P_{+i}U_{i+}\right)}{\left(P_{++}^2 - U_{++}^2\right)^2} \tag{7.19}$$

where $P_E \pm U_E$ is the observed proportion of agreement.

For kappa coefficient of agreement (κ):

For center value,

$$\kappa = \frac{(P_O - P_E)(1-P_E) - ((1-P_O-U_O)(1-P_E-U_E)U_O + U_E)U_E}{(1-P_E)^2 - U_E^2} \tag{7.20}$$

and

for uncertainty,

$$\kappa = \frac{(1-P_O-U_O)(1-P_E-U_E)(1-P_O)U_E + (1-P_E)U_O}{(1-P_E)^2 - U_E^2} \tag{7.21}$$

7.3.2 Measure of Uncertainty: Entropy

The accuracy of classification is generally measured by an error matrix. As mentioned in the previous section for assessment of soft classification, reference data should be in soft form with finer resolution. Many times this is not possible due to the non-availability of higher resolution image. Further, it is also not possible to generate fraction reference output from ground with a large number of samples. In such cases, entropy is used as an absolute measure of uncertainty (Dehghan and Ghassemian, 2006). Entropy, which is based on the information theory (Shannon 1948; Foody, 1996), can be used to estimate the uncertainty in the classification. It expresses the distribution and extent of uncertainty in a single number in information theory. Entropy of a random variable is related to the minimum attainable error probability (Feder and Merhav, 1994). Unlike the membership vector, this criterion is able to summarize the classification uncertainty in a single number per pixel, per class, or per image (Goodchild, 1995). It shows the strength of class membership assigned to particular class in the classification output.

Different forms of FERM are used to evaluate the performance of the classifier in terms of its correctness whereas RMSE and correlation coefficient are the uncertainty measures. But these methods are defined based on the difference between the expected and actual results and are relative measures. Thus, they are sensitive to error variations and not to the uncertainty variations. On the other hand, entropy calculates uncertainty from the classified data, from testing samples, without using any external data, and hence it is an indirect method to measure accuracy. Thus, entropy is an absolute measure of uncertainty, calculated only from the soft classified data without requiring any other external information. The entropy method has been used for validating the cluster formed during unsupervised clustering using FCM and IPCM (Yang and Wu, 2006).

In some of the classifiers, where membership values do not follow the probabilistic constraint, like PCM and MPCM, the entropy theorem can be utilized by rescaling (Ricotta, 2004). Thus, the average entropy (based on Shanon's entropy theorem) of the complete image can also be calculated (Dehghan and Ghassemian, 2006; Ricotta and Avena, 2006).

For a better classified output, the entropy for a known class having less uncertainty will be low, and for an unknown class with high uncertainty, it will be high in a fraction image. For example, if while taking a fraction image of a crop, the entropy value at the crop is low, the entropy value other than at the crop location will be high. Thus, low entropy means low uncertainty, which implies more accurate classified output and vice versa. A low degree of entropy (or uncertainty) means membership is associated entirely with one class and vice versa. The entropy of a classified fraction output can be computed using Equation (7.22) (Foody, 1995; Dehghan and Ghassemian, 2006):

$$\text{Entropy} = -\sum_{i=1}^{c} \mu_{ki} \log_2 \left(\mu_{ki}\right) \tag{7.22}$$

where $\log_2(\mu_{ki}) \equiv 0$ for $\mu_{ki} = 0$, c denotes the number of classes, and μ_{ki} is the estimated membership function of class i for pixel k.

For high uncertainty, i.e., low accuracy the value of entropy from Equation (7.22) is high and inverse. Entropy is defined based on actual output of classifier, so it can give the pure uncertainty of classification results (Dehghan and Ghassemian, 2006).

7.3.3 Correlation Coefficient

Correlation coefficient is used to measure the linear association between the two variables, say, X and Y. Among all the available correlation coefficients, Pearson-moment correlation coefficient is best known (DeCoursey, 2003; Kassaye, 2006). The two variables from which the correlation is to be determined from the fraction images are membership values of the classified image and membership values of the reference image. It is given by Equation (7.23).

$$r = \frac{Cov(R,C)}{\sigma^R \sigma^C} \tag{7.23}$$

where $Cov(R,C)$ represents the covariance between the reference (R) and classified (C) data. σ^R and σ^C are the standard deviations of R and C, respectively. The range of r is from −1 to +1. If the variables (R and C) are in perfect straight line, then, $r = +1$ implies increasing linear association and $r = -1$ is decreasing linear association. $r = 0$, which is a special case, shows no correlation between the variables. A value from 0.5 to 1 states a strong correlation between two variables (DeCoursey, 2003).

7.3.4 Root Mean Square Error

Root mean square error is given by taking a square root of the sum of squared difference between the membership values of the classified image and the reference image, as in Equation (7.24).

$$\text{RMSE} = \frac{\sqrt{\sum_{j=1}^{N}\sum_{i=1}^{c}\left(C_{ij} - R_{ij}\right)^2}}{M \times N} \tag{7.24}$$

where C_{ij} is the membership values from the classified image, R_{ij} is the membership values from the reference image, and $M \times N$ is the size of the image.

RMSE gives the measure of both systematic and random errors (Smith, 1997). It is an average measure of the difference of membership values of classified image to the membership values of the reference data set. The RMSE values are always greater than or equal to zero, as is evident from Equation (7.24). The interpretation of RMSE is that for good results, its value should be minimum or tend toward zero. For the given data, RMSE is calculated in two ways: global and per class. Global RMSE is the RMSE of the complete image, i.e., all the fraction images, and is given by Equation (7.24). RMSE per class can be computed using Equation (7.25).

$$\text{RMSE (class)} = \frac{\sqrt{\sum_{j=1}^{N}\left(C_{ij} - R_{ij}\right)^2}}{M \times N} \tag{7.25}$$

7.3.5 Receiver Operating Characteristic (ROC)

The receiver operating characteristic (ROC), which is based on the Neyman-Pearson detection theory, is used for the evaluation of detection performance in signal processing, communication, and medical diagnosis (Chang et al., 2001; Wang et al., 2005; Miyamoto et al., 2008; Chang, 2010). The ROC curve is used to illustrate the performance of a binary classifier system, which means whether a class is detected ('hit') or not ('miss'). The detection is measured by the area under the Neyman Pearson curve. The area is denoted by A_z and bounded between ½ and 1. For better detection, it should be closer to 1 (Wang et al., 2005). The 2D ROC curve is plotted by the false alarm rate (FAR) on one axis (x-axis) and true positive (TP) rate on another axis (y-axis). On the other hand, the 3D ROC curve is plotted by taking the false alarm rate (FAR) on the x-axis, detection threshold (t) on the y-axis, and true positive (TP) rate on the z-axis (Figure 7.1). The 2D ROC can be used for hard decision produced by the classifier, whereas 3D ROC can be used for the soft decision (Wang et al., 2005).

This method is also able to check the accuracy while extracting single land cover from remote sensing image classification, where the classifier acts like a binary or in other words, when the interest is only to know whether the classifier is able to detect a particular class or not. The 2D ROC curve is plotted by the true positive (TP) rate on one axis and the false alarm rate (FAR) on the other axis, whereas in 3D ROC

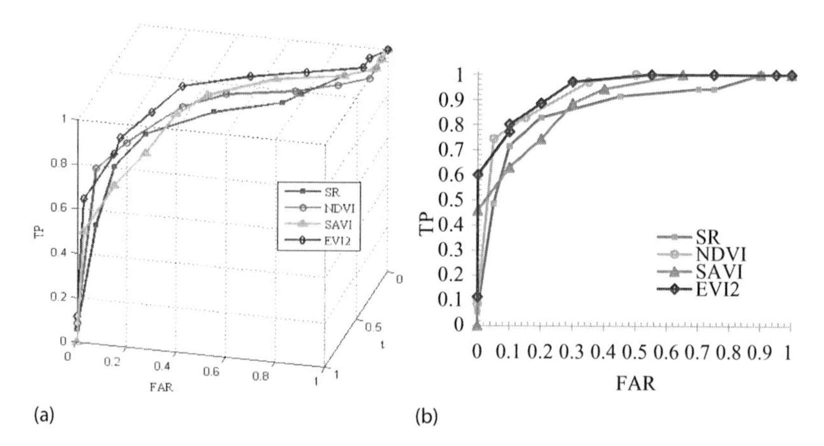

FIGURE 7.1 ROC curves for identification of wheat crop using noise clustering classifier. (a) 3-D curve showing variation of true positive (TP) with false alarm rate (FAR) and detection threshold (t) and (b) variation of TP with FAR alone.

one more axis corresponding to the detection threshold is also added to the 2D ROC curves. The TP and FAR can be defined in Equations (7.26) and (7.27), respectively:

$$\text{TP} = \frac{\text{Total number of target pixels detected as target}}{\text{Total number of target pixels present in the sample}} \tag{7.26}$$

$$\text{FAR} = \frac{\text{Total number of background pixels detected as target}}{\text{Total number of background pixels present in the sample}} \tag{7.27}$$

The area under the 2D ROC curve is used to measure the detection performance of the classifier. It is bounded between ½ and 1. For better detection, it should be closer to 1.

7.3.6 Method for Edge Preservation

There are methods for adding contextual information through MRF or local convolution to reduce noisy pixels in a given image. While adding contextual information, edges of the objects are affected due to smoothing effect. Therefore, it is important to verify whether any edge within a classified output is correct or not. An edge represents boundaries between two objects which may be characterized as a step function or slope between two regions (Wen and Xia, 1999).

As per Wen and Xia (1999), if for some specific threshold c,

$|\mu_1 - \mu_2| \leq c$ then there is no significant difference between the grey levels on the two sides of the edge whereas if $|\mu_1 - \mu_2| > c$, there will be a significant difference between the true averages, where μ_1 and μ_2 are the mean value of the pixels on each side of the edges.

To verify the significance of an edge, the distribution of grey levels of both sides needs to be analyzed in the sense that the difference between the average values within two regions represents the steepness of the edge. To determine the value of c, edge point is examined first through Equation (7.28).

$$|X_i - Y_i| > c + \sqrt{2}SZ_\alpha \tag{7.28}$$

where X_i and Y_i represent the grey level of ith pixel on two sides of the edge respectively. S is the standard deviation of the grey levels in the region the point. Z_α can be obtained from the standard distribution tables. In practice, the values of α can be assigned as 0.01, 0.05, 0.1, 0.2 depending upon different requirements. Both low and high thresholds for an edge can be identified by selecting two different values.

The fraction image generated using the contextual approach for a particular class has the high membership value if class exists for a known location. If it does not exist, then membership value is low. For a homogeneous area, the fraction image will have less variability among the membership values. Consequently, the mean membership value will be high and variance will be low for a homogeneous area. This concept has been suggested to verify the edge preservation.

For edge preservation method, first a homogeneous area of a specific class, i.e., crop, has been selected which has a high mean value and a low variance. After selecting a homogeneous area, two sets of pixels were selected at either side of the crop class edge. Mean and variance are calculated for these two sets of pixels in each iteration. The mean difference of these two sets of pixels should be high and variance within should be low if the edge is to be preserved.

7.4 SUMMARY

In this chapter, focus has been kept on assessment of accuracy of soft classified output fraction images. In literature, methods have been developed for assessment of accuracy of soft output. Relative assessments from finer reference images to absolute indirect assessment methods of soft classified outputs have been explained in this chapter. Relative methods were image-to-image accuracy using FERM using single operator as well as composite operators such as MIN-MIN, MIN-LEAST, MIN-PROD, and SCM. Absolute indirect accuracy assessment methods include entropy, correlation coefficient, RMSE, and ROC.

BIBLIOGRAPHY

Binaghi, E., Brivio, P.A., Chessi, P. and Rampini, A., 1999. A fuzzy set based accuracy assessment of soft classification. *Pattern Recognition Letters*, 20, 935–948.

Carpenter, G.A., Gopal, S., Martens, S. and Woodcock, C.E., 1999. A neural network method for mixture estimation for vegetation mapping. *Remote Sensing of Environment*, 70, 138–152.

Chang, C.I., 2010. Multi parameter receiver operating characteristic analysis for signal detection and classification. *IEEE Sensors Journal*, 10 (3), 423–442.

Chang, C.I., Ren, H., Chiang, S.S. and Ifarraguerri, A., 2001. An ROC analysis for subpixel detection. *IEEE 2001 International Geoscience and Remote Sensing Symposium*, Australia, July 24–28.

Chen, J., Zhu, X., Imura, H. and Chen, X., 2010. Consistency of accuracy assessment indices for soft classification: Simulation analysis. *ISPRS Journal of Photogrammetry and Remote Sensing*, 65, 156–164.

Congalton, R.G., 1991. A review of assessing the accuracy of classifications of remotely sensed data. *Remote Sensing of Environment*, 37, 35–47.

Congalton, R.G. and Green, K., 1999. *Assessing the Accuracy of Remotely Sensed Data: Principles and Practices*. Boca Raton, FL: Lewis.

Cross, A.M., Settle, J.J., Drake, N.A. and Paivinen, R.T.M., 1991. Sub-pixel measures of tropical forest cover using AVHRR data. *International Journal of Remote Sensing*, 12, 1119–1129.

DeCoursey, W.J., 2003. *Statistics and Probability for Engineering Applications*. Newnes, Australia: Elsevier Science.

Dehghan, H. and Ghassemian, H., 2006. Measurement of uncertainty by the entropy: Application to the classification of MSS data. *International Journal of Remote Sensing*, 27 (18), 4005–4014.

Feder, M. and Merhav, N., 1994. Relations between entropy and error probability. *IEEE Transaction of Information Theory*, 40 (1), 259–266.

Fisher, P.F. and Pathirana, S., 1990. The evaluation of fuzzy membership of land cover classes in the suburban zone. *Remote Sensing of Environment*, 34, 121–132.

Foody, G.M., 1995. Cross entropy for the evaluation of the accuracy of a fuzzy land cover classification with fuzzy ground data. *ISPRS Journal of Photogrammetry and Remote Sensing*, 50, 2–12.

Foody, G.M., 1996. Approaches for the production and evaluation of fuzzy land cover classifications from remotely-sensed data. *International Journal of Remote Sensing*, 17 (7), 1317–1340.

Foody, G.M., 2002. Status of land cover classification accuracy assessment. *Remote Sensing of Environment*, 80, 185–201.

Goodchild, M.F., 1995. Attribute accuracy. In: Guptil S.C. and Morrison J.L. (eds.), *Elements of Spatial Data Quality*. Oxford: Elsevier Scientific, Oxford, pp. 59–79.

Gopal, S. and Woodcock, C., 1994. Theory and methods for accuracy assessment of thematic maps using fuzzy sets. *Photogrammetric Engineering and Remote Sensing*, 60, 181–188.

Gutman, G. and Ignatov, A., 1998. The derivation of the green vegetation fraction from NOAA/AVHRR data for use in numerical weather prediction models. *International Journal of Remote Sensing*, 60 (2), 181–188.

Jensen, J.R., 1996. *Introductory Digital Image Processing: A Remote Sensing Perspective*. 2nd Edition. Prentice Hall PTR, ISBN: 978-0131453616.

Kassaye, R.H., 2006. Suitability of Markov random field-based method for super-resolution mapping, M.Sc. Thesis, GFM, International Institute of Geoinformation Science and Earth Observation.

Kumar, A. and Dadhwal, V.K., 2010. Entropy-based fuzzy classification parameter optimization using uncertainty variation across spatial resolution. *Journal of the Indian Society of Remote Sensing*, 38 (2), 179–192.

Kuzera, K. and Pontius Jr., R., 2004. Categorical coefficients for assessing soft-classified maps at multiple resolutions. In: *Conference Proceedings of the Joint Meeting of the Fifteenth Annual Conference of the International Environmetrics Society and the Sixth Annual Symposium on Spatial Accuracy Assessment in Natural Resources and Environmental Sciences*, vol. 28. Portland, ME.

Latifovic, R. and Olthof, I., 2004. Accuracy assessment using sub-pixel fractional error matrices of global land cover products derived from satellite data. *Remote Sensing of Environment*, 90, 153–165.

Lewis, H.G. and Brown, M., 2002. A generalized confusion matrix for assessing area estimates from remotely sensed data. *International Journal of Remote Sensing*, 22, 3223–3235.

Liu, W. and Wu, E.Y., 2005. Comparison of nonlinear mixture models: Sub-pixel classification. *Remote Sensing of Environment*, 94, 145–154.

Meyer, M. and Werth, L., 1990. Satellite data: Management panacea or potential problem? *Journal of Forestry*, 88 (9), 10–13.

Miyamoto, S., Ichihashi, H. and Honda, K., 2008. *Algorithms for Fuzzy Clustering, Studies in Fuzziness and Soft Computing*, Berlin, Heidalberg: Springer-verlag, 229, 65–66.

Okeke, F. and Karnieli, A., 2006. Methods for fuzzy classification and accuracy assessment of historical aerial photographs for vegetation change analyses. Part I: Algorithm development. *International Journal of Remote Sensing*, 27 (1), 153–176.

Ozdogan, M. and Woodcock, C.E., 2006. Resolution dependent errors in remote sensing of cultivated areas. *Remote Sensing of Environment*, 103, 203–217.

Pontius Jr., R.G. and Cheuk, M.L., 2006. A generalized cross tabulation matrix to compare soft classified maps at multiple resolutions. *International Journal of Geographical Information Science*, 20 (1), 1–30.

Pontius Jr., R.G. and Connors, J., 2006. Expanding the conceptual, mathematical and practical methods for map comparison. In: *Conference Proceedings of the Meeting of Spatial Accuracy 2006*, Lisbon, Portugal, pp. 64–79.

Ricotta, C., 2004. Evaluating the classification accuracy of fuzzy thematic maps with a simple parametric measure. *International Journal of Remote Sensing*, 25 (11), 2169–2176.

Ricotta, C. and Avena, G.C., 2006. Evaluating the degree of fuzziness of thematic maps with a generalized entropy function: A methodological outlook. *International Journal of Remote Sensing*, 23 (20), 4519–4523.

Shabanov, N.V., Lo, K., Gopal, S. and Myneni, R.B., 2005. Subpixel burn detection in moderate resolution imaging spectro-radiometer 500-m data with ARTMAP neural networks. *Journal of Geophysical Research*, 110, 1–17.

Shannon, C.E., 1948. A mathematical theory of communication. *AT&T Technical Journal*, 27, 379–423.

Silván-Cárdenas, J.L. and Wang, L., 2008. Sub-pixel confusion–uncertainty matrix for assessing soft classifications. *Remote Sensing of Environment*, 112 (3), 1081–1095.

Smith, S.W., 1997. *DSP Guide for Scientists and Engineers*, Chapter 2, San Diego, CA: California Technical.

Stehman, S.V. and Czaplewski, R.L., 1998. Design and analysis for thematic map accuracy assessment: Fundamental principles. *Remote Sensing of Environment*, 64, 331–344.

Tortora, R., 1978. A note on sample size estimation for multinomial populations. *The American Statistician*, 32 (3), 100–102.

Townsend, P.A., 2000. A quantitative fuzzy approach to assess mapped vegetation classifications for ecological applications. *Remote Sensing of Environment*, 72, 253–267.

Wang, J., Chang, C.I., Yang, S.-C., Hsu, G.C., Hsu, H.H., Chung, P.-C., Guo, S.M., Lee, S.K., 2005. 3D ROC analysis for medical diagnosis evaluation. *Proceedings of the 27th Annual International Conference of the IEEE Engineering in Medicine and Biology Society (EMBS)*. Shanghai, China, September 2005, pp. 7545–7548.

Wen, W. and Xia, A., 1999. Verifying edges for visual inspection purposes. *Pattern Recognition Letters*, 20 (3), 153–172.

Xu, M., Cao, C., Zhang, H., Guo, J., Nakane, K., He, Q., Guo, J. et al., 2010. Change detection of an earthquake-induced barrier lake based on remote sensing image classification. *International Journal of Remote Sensing*, 31, 3521–3534.

Yang, M. and Wu, K., 2006. Unsupervised possibilistic clustering. *Pattern Recognition*, 39, 5–21.

....Begin your day with....

Gratitude....

With Positive Vision....

Have Broader Plans....

Do Small Things Right Way....

Help Someone....

Appendix: A1, SMIC: Sub-Pixel Multi-Spectral Image Classifier Package

This appendix: A1 provides an in-depth discussion on the various capabilities of *SMIC: sub-pixel multi-spectral image classifier* package. The commercially available software has limitations for various classifying algorithms such as statistical based linear mixture model (LMM), fuzzy set based fuzzy *c*-means (FCM), and artificial neural network (ANN). Limited algorithms have been incorporated in different commercially available digital image processing software like neural network and unsupervised fuzzy *c*-Mean in PCI Geomatica and linear mixture model (LMM) in ERDAS, ENVI, etc.

However, the SMIC package is menu driven type with a large number of fuzzy based supervised soft classifiers. There is also an option to add contextual information through MRF or by local convolution based methods. Outputs from the SMIC package can be generated as soft or hard classified outputs and all parameters are open to do the experiment with different values assigned to all variables in fuzzy based algorithm. Fuzzy based algorithms discussed in Chapters 3, 5, 6, and 7 have been implemented through in-house developed package called SMIC (Figure A1.1).

This package has three modules. The first module has the capability to select the distance as well as kernel based fuzzy classifiers. Through Figure A1.2a, various distances can be applied in a fuzzy based classifier, and through Figure A1.2b, various kernels can be applied in chosen fuzzy based classifier. It also has the provision to add contextual information through Markov random field and the local convolution methods. While classifying single date image, the classification GUI can be opened up and the fuzzy classifiers can be selected with an option to add the contextual part through MRF mode or local convolution mode. The soft output membership values generated using different classifiers are saved as fraction images.

The second module is developed to process temporal data sets for specific class extraction (Figure A1.3).

In this module, the spectral dimensionality of data is reduced through the class based sensor dependent (CBSI) approach (Chapter 6, Section 6.2), and the temporal dimensionality can be maintained. Figure A1.4a provides an option to invoke the CBSI indices palate whereas Figure A1.4b provides an option for selecting mathematical indices expression while processing the temporal data. The advantage of CBSI is that the user does not have to specify band information from image data sets.

Another main step in this package is to generate training sample data through region growing methods (Figure A1.5). In the SMIC package, the region growing option has been implemented. This region growing method collects similar pixels

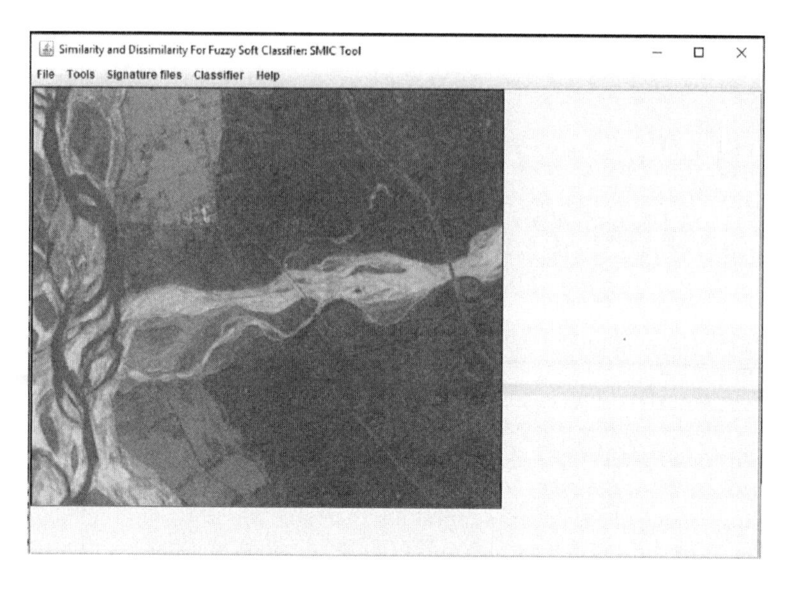

FIGURE A1.1 The main window of the SMIC package.

as training data sets using seed pixels while having various similarity/dissimilarity options. It also generates class based sensor independence indices using seed pixels from temporal images.

The third module in SMIC is for the assessment of accuracy of soft classified data through the fuzzy error matrix (FERM) (Chapter 7, Section 7.3.1.1) using soft reference data (Figure A1.6). For sub-pixel classification, when reference data and the outputs are soft, FERM and other modified versions of it can be generated using image-to-image accuracy of soft classified output. Here the requirement of assessment of accuracy of soft classified data is soft reference data. So this package provides assessment of accuracy of soft classified data as image-to-image assessment of accuracy.

The SMIC has been developed using JAVA programming language and is specifically capable of soft/hard land cover mapping from remote sensing multi-spectral data. It has the capability to process the multi-spectral mono/temporal remote sensing satellite data at a sub-pixel level with various fuzzy based classifiers. The system can handle multi-spectral images of any number of bands. In this package, various classification algorithms such as fuzzy *c*-means, possibilistic *c*-means, noise clustering, modified possibilistic *c*-means, and improved possibilistic *c*-means with or without entropy have been incorporated in supervised mode with soft/hard classification options. These algorithms can also be used with MRF or local convolution methods to add contextual information to reduce noise.

Reference training data can be generated from this system in two modes; one is manually and other is region growing concept. In manual mode, pixel-by-pixel training samples can be generated (Figure A1.7). In manual mode the chances of picking non-homogeneous training samples are higher, hence in such cases region growing through seed pixel can be used for collecting training samples (Figure A1.5). Additionally, in region growing similarity/dissimilarity algorithms have been

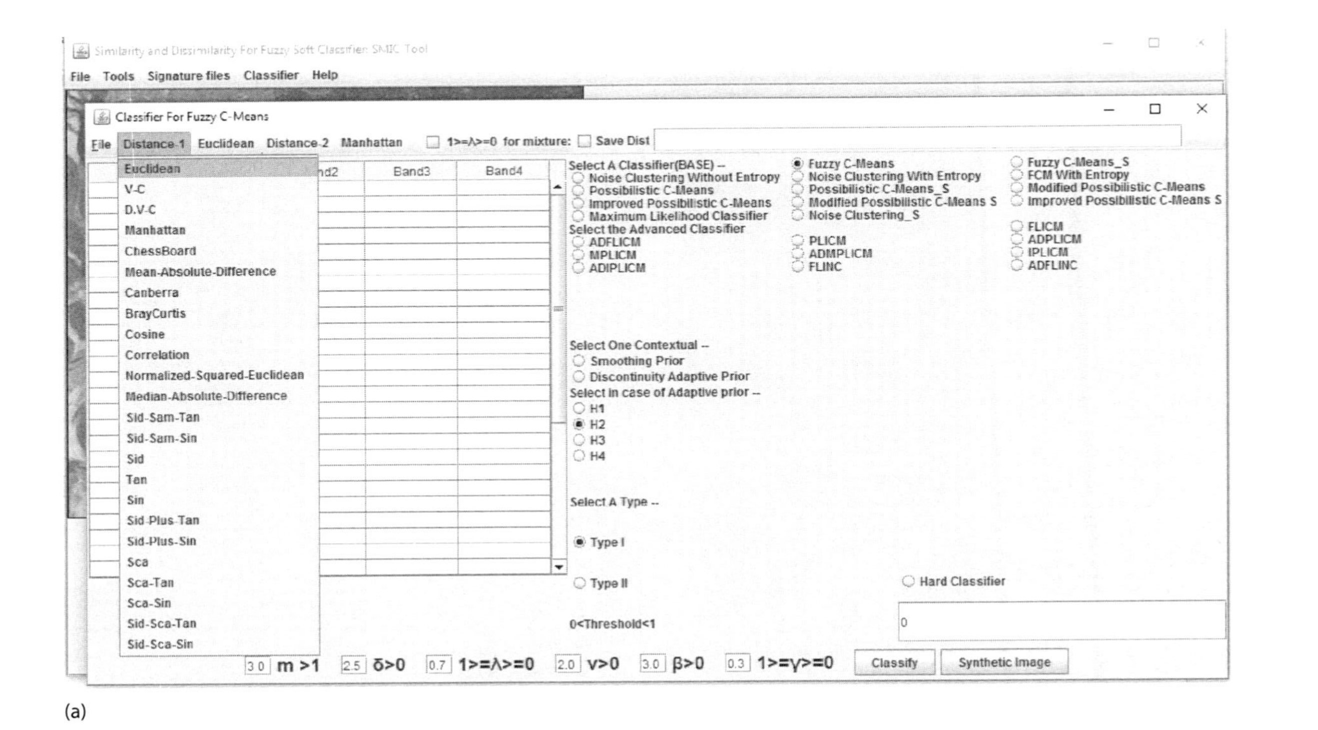

(a)

FIGURE A1.2 (a) SMIC with distance norm.

(Continued)

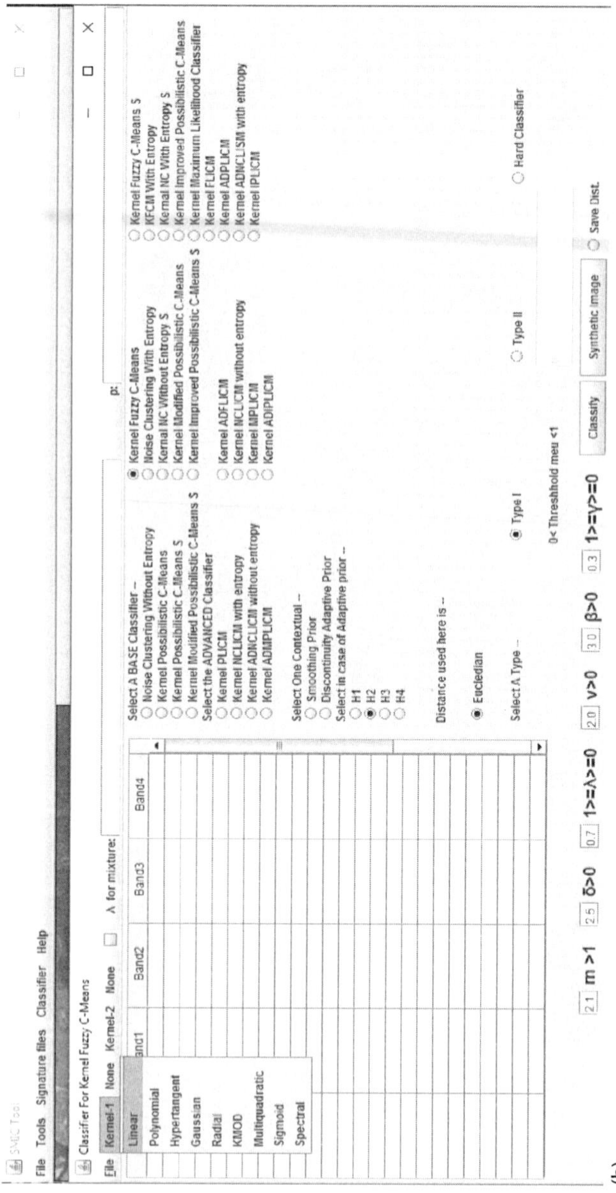

SMIC Tool
File Tools Signature files Classifier Help
Classifier For Kernel Fuzzy C-Means
File Kernel-1 None Kernel-2 None Λ for mixture: p:
Linear
Polynomial
Hypertangent
Gaussian
Radial
KMOD
Multiquadratic
Sigmoid
Spectral
Band1 Band2 Band3 Band4
Select A BASE Classifier --
Noise Clustering Without Entropy
Kernel Possibilistic C-Means
Kernel Possibilistic C-Means S
Kernel Modified Possibilistic C-Means S
Kernel Fuzzy C-Means
Noise Clustering With Entropy
Kernal NC Without Entropy S
Kernel Modified Possibilistic C-Means
Kernel Improved Possibilistic C-Means S
Kernel Fuzzy C-Means S
KFCM With Entropy
Kernal NC With Entropy S
Kernel Improved Possibilistic C-Means
Kernel Maximum Likelihood Classifier
Select the ADVANCED Classifier
Kernel PLICM
Kernel NCLICM with entropy
Kernel ADNCLICM without entropy
Kernel ADMPLICM
Kernel ADFLICM
Kernel NCLICM without entropy
Kernel MPLICM
Kernel ADIPLICM
Kernel FLICM
Kernel ADPLICM
Kernel ADNCLISM with entropy
Kernel IPLICM
Select One Contextual --
Smoothing Prior
Discontinuity Adaptive Prior
Select in case of Adaptive prior --
H1
H2
H3
H4
Distance used here is --
Eucledian
Select A Type -- Type I Type II Hard Classifier
0< Threshhold meu <1
2.1 m >1 2.5 δ>0 0.7 1>=λ>=0 2.0 v>0 3.0 β>0 0.3 1>=γ>=0 Classify Synthetic image Save Dist.

(b)

FIGURE A1.2 (Continued) (b) SMIC with kernel norms.

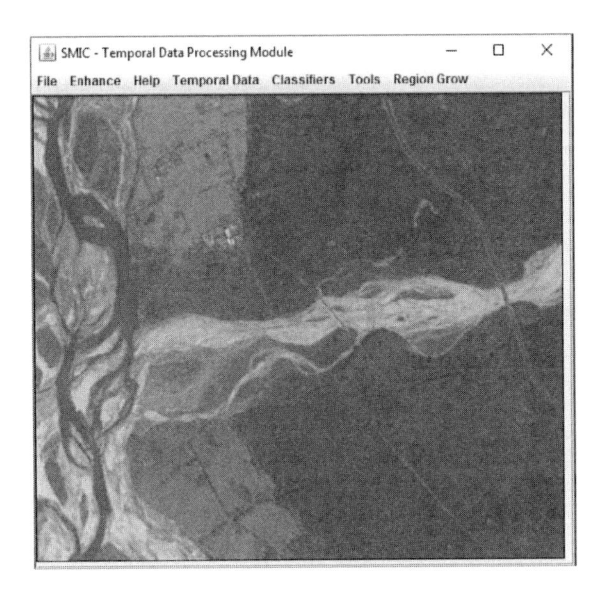

FIGURE A1.3 Temporal data processing module.

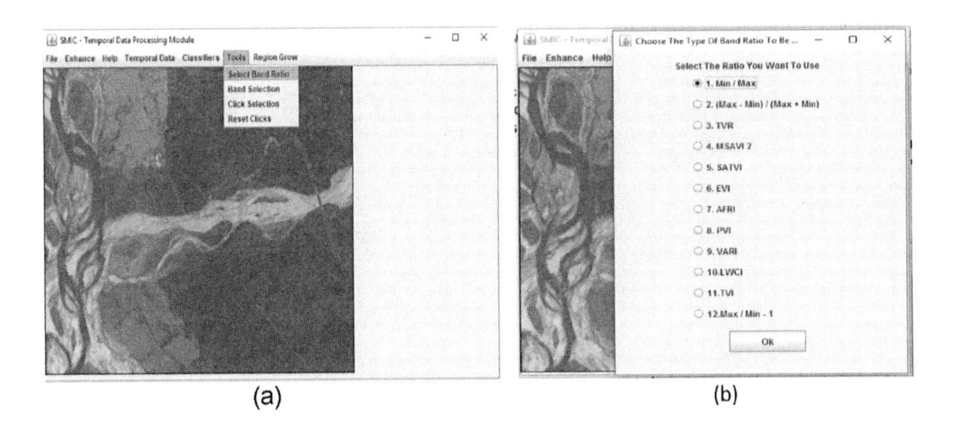

FIGURE A1.4 (a) and (b) Dimensionality reduction methods.

implemented for generating the criteria to pick similar pixels in a given search window. Training sample data of each class is saved in separate training files.

Minimum human intervention has been considered while developing the module for processing temporal data. This module includes temporal information as temporal indices data while reducing spectral information of multi-spectral images. The advantage of including temporal data from temporal images is that, it provides unique information about specific class mapping due to reduced spectral overlapping with other classes. Main GUI of SMIC temporal data processing package has been given in Figure (A1.3). Figure A1.8 shows how to load other temporal images.

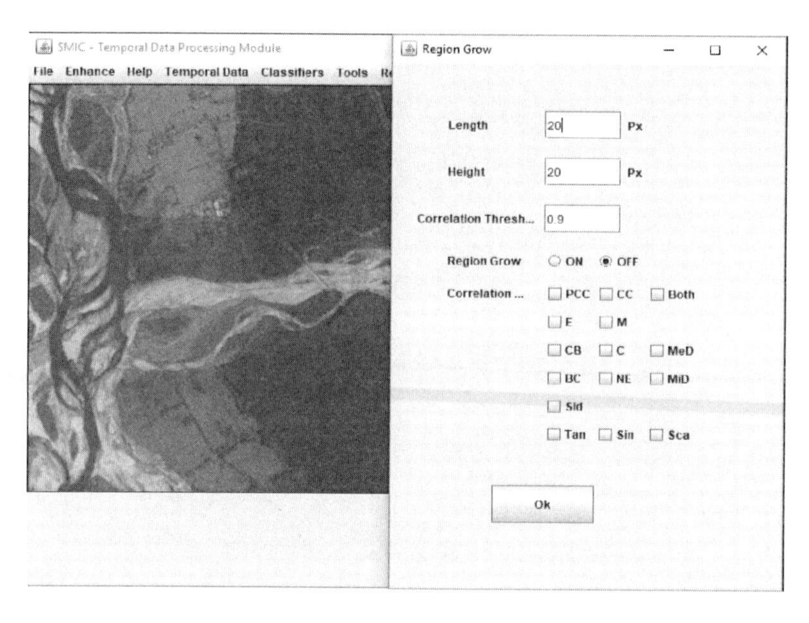

FIGURE A1.5 Region growing methods.

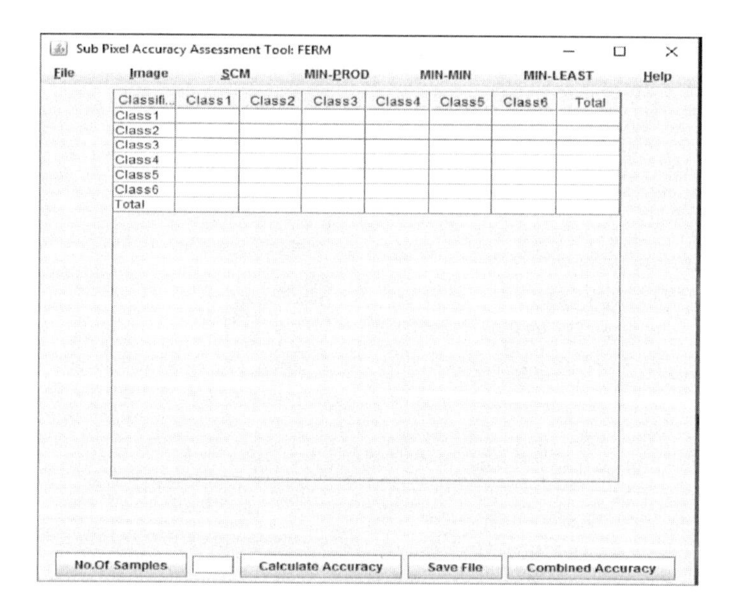

FIGURE A1.6 FERM tool for assessment of accuracy.

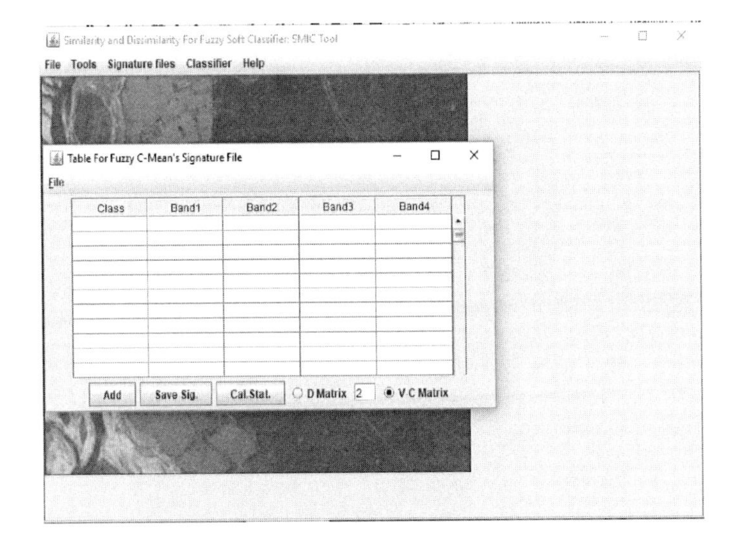

FIGURE A1.7 Pure signature data collection module.

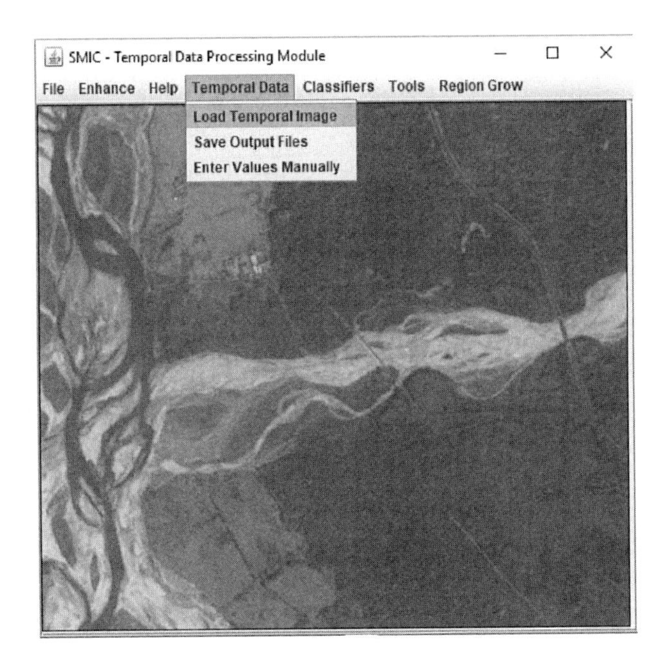

FIGURE A1.8 Load other temporal images.

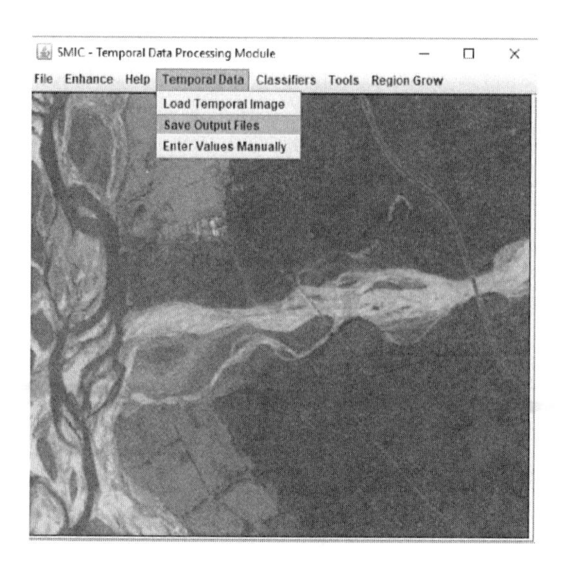

FIGURE A1.9 Saving indices outputs.

Figure A1.9 provides an option to save indices outputs generated. These indices layers generated from temporal data sets give information as a temporal indices database.

After generating the temporal indices outputs, the next step is to apply a fuzzy based classifier capable to identify a specific class of interest. In this module, PCM, NC, and MPCM classifiers (Chapter 3) have been implemented with various similarity/dissimilarity measures (Figure A1.10). The advantage of applying these classifiers is due to algorithms' capability to identify the single class of interest with minimum parameters.

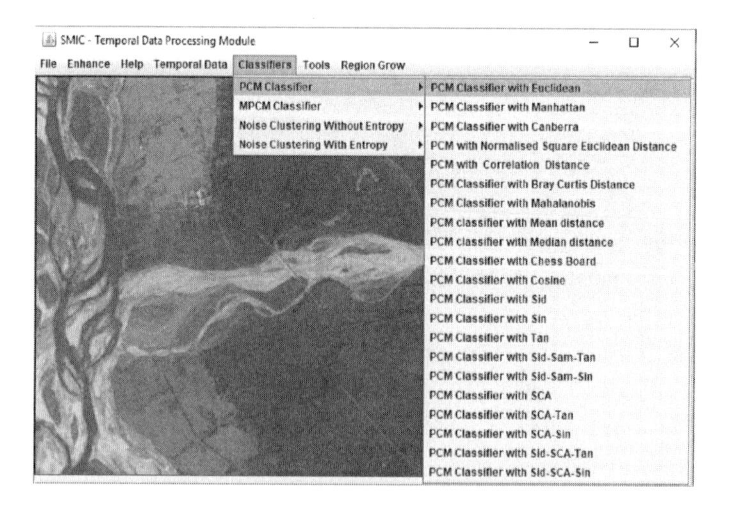

FIGURE A1.10 Classification methods for specific class mapping.

Another advantage of using the SMIC module for temporal data processing is that the user needs just four steps to identify a specific class of interest using temporal data. Many times it is not possible to get temporal multi-spectral data from the same remote sensing sensor. In that case, this package has the capability to use the remote sensing multi-spectral data available from different sensors. In addition, this module does not require knowledge about different spectral bands present in remote sensing multi-spectral images. This module applies statistical operators to identify suitable bands to be used in indices generation to have a temporal indices database. Therefore, the overall advantage of this module is that technological knowledge is on the back end, and it requires minimum knowledge to work with temporal datasets for specific class identification.

While mapping a specific crop, for example, the knowledge of other crops present in that area is also important. Further, it is a fact that the spectral responses of different crops are likely similar to each other on a particular date and make the process complex for mapping specific crops using single date imagery. Temporal information of a crop can, however, provide a good solution for discriminating among various crops and vegetation classes using the differences in their growth patterns as a discriminating factor. The need of temporal data for continuous monitoring of crops and the unavailability of continuous temporal data is also a well-known problem. Hence a multi-sensor approach for increasing the temporal data sampling for monitoring crops has to be evaluated for its effectiveness. SMIC is capable to handle the mono as well as multi-sensor temporal images and hence an effective package for specific class identification and for mixed pixel handling.

See Inside Shows Blessed With....

See Outside Shows Missing Out....

Appendix: A2, Case Studies from SMIC Package

CASE STUDY 1: STUDY OF SIMILARITY AND DISSIMILARITY MEASURES WITH IPCM AND MPCM CLASSIFIERS

As mentioned earlier, in remote sensing images there are both pure and mixed pixels. Hard classifiers are found to be inappropriate to address this mixed pixel problem. Fuzzy set theory and possibilistic based algorithms such as fuzzy c-means (FCM), possibilistic c-means (PCM), etc. account for information of classes mixed within a pixel, and hence are quite successful in dealing with the mixed pixel problem. In the literature, PCM classifiers have only been studied with Euclidean, Mahalanobis, and diagonal Mahalanobis norms, but other similarity and dissimilarity functions have not been applied with it.

In this case study, possibilistic based fuzzy algorithms such as possibilistic c-means (PCM), improved possibilistic c-means (IPCM), and modified possibilistic c-means (MPCM) have been studied with twelve similarity and dissimilarity measures: cosine, correlation, Euclidean, Manhattan, Bray Curtis, Canberra, chessboard, mean absolute difference, median absolute difference, variance–covariance, diagonal variance covariance, and normalized squared Euclidean in both single and composite mode. Testing and implementation of norms have been performed on the simulated Formosat-2 Image with five different classes (viz., water, wheat, forest, riverine sand, and fallow land) in different proportions.

The performance of all similarity and dissimilarity norms with fuzzy based algorithms have been tested on the simulated image for their capability of detecting different class pixels from mixing regions of varying class proportions and for their strength of suppressing within class variance. Optimization of performance of norms has been done with respect to varying weighted exponent parameter (m). Based on the results obtained from simulated images, the optimized parameters have been applied on real Formosat-2 image (8m multispectral). The effect of number of classes classified by the similarity and dissimilarity measures using PCM, IPCM, and MPCM classification results have also been studied.

OBJECTIVES OF THE CASE STUDY

In these case studies, the following research objectives are framed.

1. To study various similarity and dissimilarity measures with PCM, IPCM, and MPCM (Chapter 3).
2. To study the capability of all similarity and dissimilarity norms with PCM, IPCM, and MPCM for the suppression of within class variance (interclass variance) and detection of mixed pixels in the double as well as triple class mixing region (with varying mixing regions).

3. To perform the comparative analysis of performance of all similarity and dissimilarity norms with PCM, IPCM, and MPCM.
4. To optimize the similarity and dissimilarity measures with PCM, IPCM, and MPCM with respect to weighted exponent parameter ("m") while extracting one or more classes.

Study Area and Data Used

The study area is located toward the east side of Haridwar city, district Haridwar, Uttarakhand, India. In state Uttarakhand, it shares boundaries with districts Dehradun in the northeast and Pauri Garhwal to the east, whereas with Uttar Pradesh it shares boundaries with Muzaffarnagar and Bijnor in the south and Saharanpur in the west.

The central latitude and longitude of the city are 29.956°N and 78.170°E, respectively. The location map and different land cover classes identified in the study area are shown in Figure A2.1. Five types of land cover classes, water, wheat, forest, riverine sand and fallow land, have been identified in the study area. The reasons for selecting this area are diversity in terms of land use classes such as vegetation types, sand/clay, dense forest, fallow agriculture land, and water as well as field data and availability of satellite data.

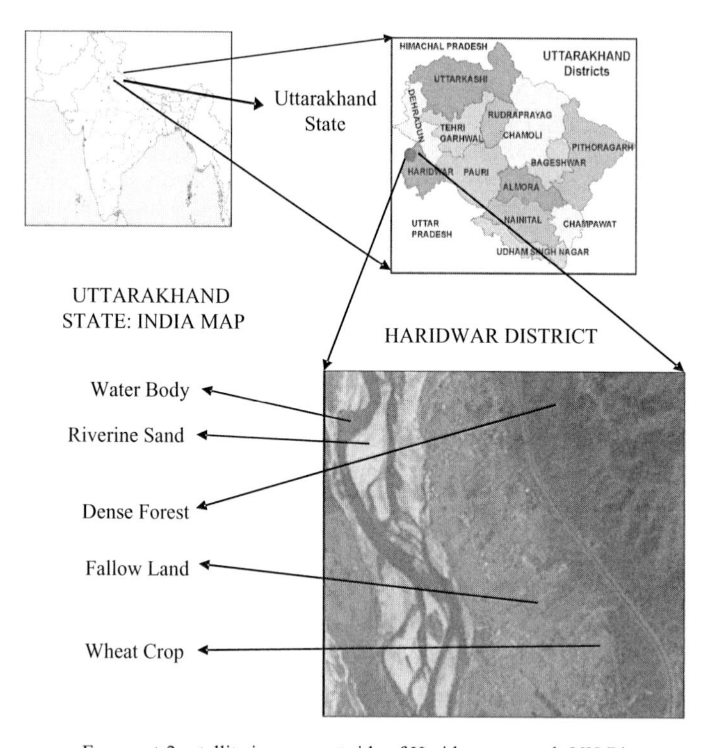

FIGURE A2.1 The location map and land cover classes identified in the study area.

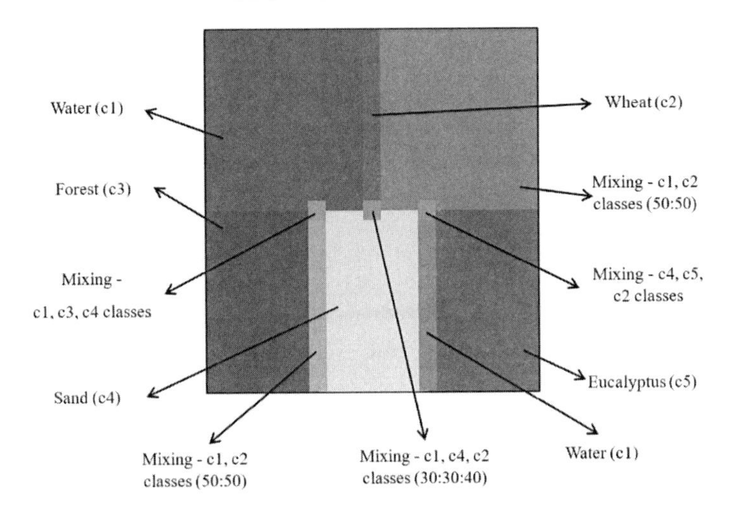

FIGURE A2.2 Simulated image generated from Formosat-2 image.

In this study, multi-spectral images and simulated images of Formosat-2 (Figure A2.2) satellite were used. The main data set that has been used for image processing aspect is simulated Formosat-2 image. Formosat-2 is the first remote sensing satellite developed by National Space Organization (NSPO), Taiwan, and was launched on May 21, 2004. The main aim of the Formosat-2 mission has been to capture remotely sensed data on land and oceans of the earth with a daily revisit. The Formosat-2 satellite carries both "remote sensing" and "scientific observation" tasks in its mission. It supports monitoring and detecting land change for any specific regions for various industries and mapping applications. Table A2.1 shows the Formosat-2 sensor specifications.

TABLE A2.1
Formosat-2 Satellite Sensor Specifications

Spectral bands (Spectral resolution)	P: 0.45–0.90 μm (Panchromatic)
	B1: 0.45–0.52 μm (Blue)
	B2: 0.52–0.60 μm (Green)
	B3: 0.63–0.69 μm (Red)
	B4: 0.76–0.90 μm (Near-infrared)
Sensor footprint	24 × 24 km
Revisit interval	Daily
Viewing angles	Cross track and along track (forward/aft): ± 45 degree
Image dynamics	8 bits per pixel
Resolution	PAN (Panchromatic): 2 m
	MS (Multi-spectral): 8 m

Methodology

Adopted methodology for the overall analysis and implementation of algorithms with all similarity and dissimilarity norms is shown in the form of a flow diagram in Figure A2.3. A simulated Formosat-2 image has been used to test the implemented norms. The training data files for each class (water, wheat, forest, riverine sand, and fallow agriculture land) has been taken from the simulated Formosat-2 image. On the simulated image, the classification experiments have been performed using PCM, IPCM, and MPCM and it has been applied by integrating all the similarity

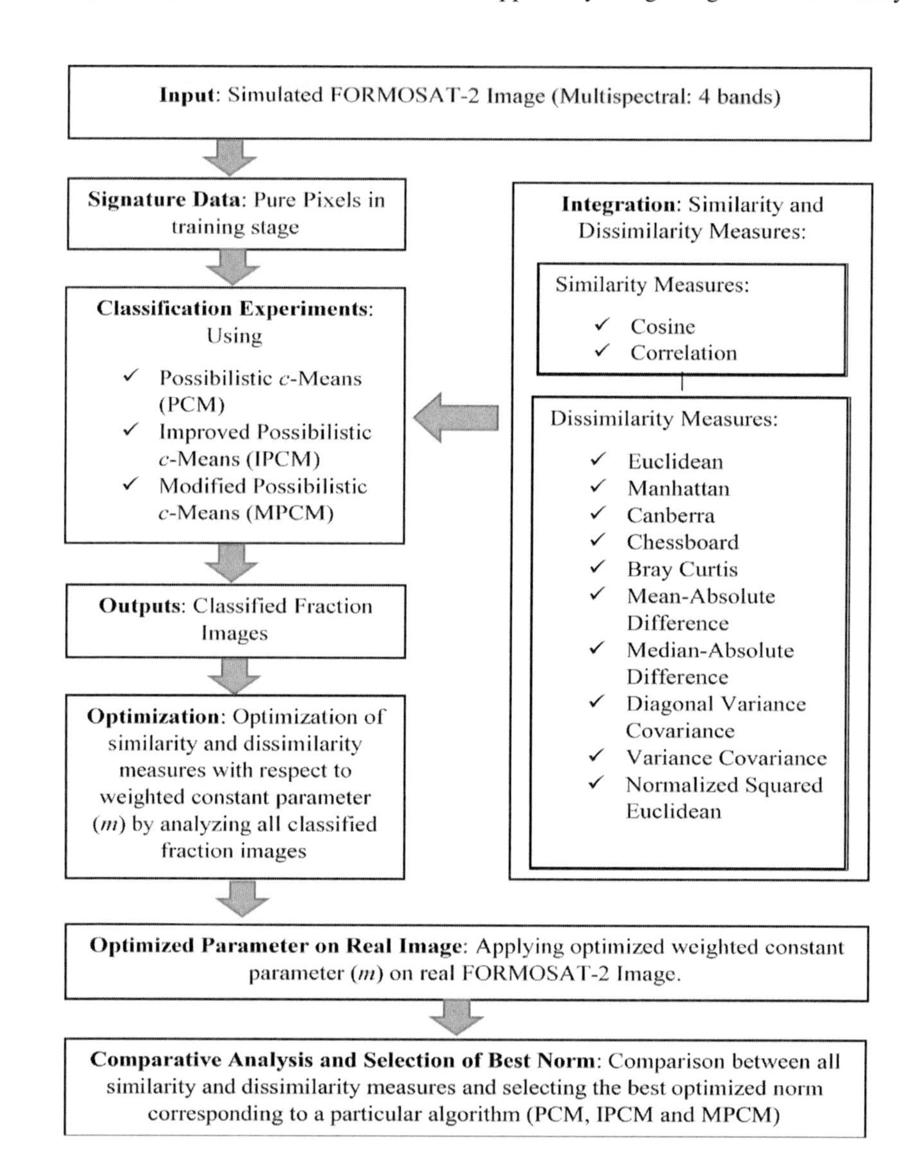

FIGURE A2.3 Methodology adopted for Case Study 1.

and dissimilarity measures one by one. These classification experiments have been conducted in five different cases while considering a particular algorithm:

Case 1: Two class classification: When any two classes have been classified (Water and Wheat).
Case 2: Three classes classification: When any three classes have been classified (Water, Wheat, and Forest).
Case 3: Four classes classification: When any four classes have been classified (Water, Wheat, Forest, and Riverine Sand).
Case 4: All five classes classification: When all five classes have been classified (Water, Wheat, Forest, Riverine Sand, and Fallow Agriculture Land).

Results

Results showed that, in the case of PCM, Euclidean ($m = 3$) and cosine ($m = 3$) (Figure A2.4; two classes classification case results: Water and Wheat classified) were found to be the best norms when applied for lower number of classes (two or three), whereas the normalized squared Euclidean Distatnce (ED) was the best norm when the number of classes are increased. In the case of IPCM also, the ED was the best norm when used for two or three classes, whereas for a large number of classes, increased variance covariance performed well at $m = 1.7$ and 2.5. Normalized squared ED has shown special affinity toward MPCM algorithm, hence it works perfectly with MPCM at extreme values, $m = 1.1$ and 3.0. Algorithms have shown specific behavior to a particular norm. Thus, apart from the conventional Euclidean norm, other norms like cosine, normalized squared Euclidean, and variance–covariance, have performed nicely with PCM, IPCM, and MPCM algorithms. The study therefore signifies that the integration of similarity and dissimilarity measures with possibilistic based fuzzy algorithms is an effective approach for dealing with sub-pixel issues.

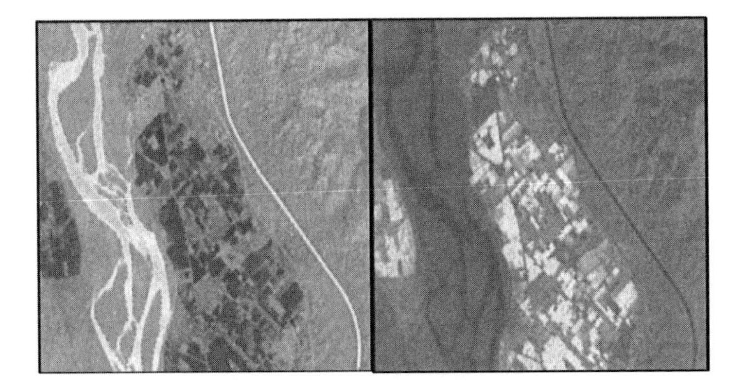

FIGURE A2.4 Fraction images generated by cosine at $m = 3.0$.

CASE STUDY 2: BI-SENSOR TEMPORAL DATA FOR PADDY CROP MAPPING

Use of single date imagery of optical remote sensing can create problems in the discrimination of specific crops. This problem can be resolved by using a temporal dataset, but issues like cloud cover and atmospheric disturbances can create gaps in the temporal optical data. To overcome these problems of cloudy images, microwave data is preferred because it has the ability to penetrate through the atmosphere and clouds. These issues are considered and resolved while integrating RISAT-1 and Formosat-2 temporal images for the identification of late transplant paddy fields and harvested fields. Three datasets in different temporal combinations of microwave and optical images are considered. Another challenge is handling the mixed pixels, which is resolved using fuzzy classifiers. In this study, IPCM and MPCM algorithms (Chapter 3) are used with different similarity and dissimilarity measures for the identification of a specific crop, i.e., paddy field.

OBJECTIVES OF THE CASE STUDY

The following research objectives are framed in this case study.

1. To study the effectiveness of bi-sensor temporal data for paddy field identification.
2. To study the capability of IPCM and MPCM for single class extraction (Chapter 6).
3. To study various norms with IPCM and MPCM classifiers.

STUDY AREA AND DATA USED

The geographical boundaries of the study area are identical to Case Study1. The holy Ganga river flows through the study area, and the land is fertile and rich for agriculture purposes. Major crops cultivated in this region are paddy, mustard, wheat, sugarcane, groundnuts, and fruits like mangoes and litchis. The study area is 12.3 km^2. The temperature in summer ranges from 23°C to 46°C while in winter from 0°C to 25°C. Figure A2.5 shows the details of existing land cover in the study area.

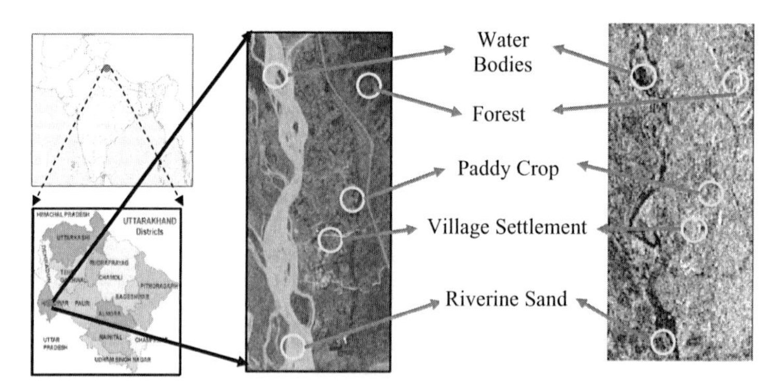

FIGURE A2.5 Study area and details of land cover in optical and RiSAT-1 data.

Many studies have been conducted for crop mapping using the temporal optical data. These studies were conducted generally on the phenology of the crop. However, in India the monsoon season coincides with the transplant season for the paddy crop. Therefore, in this case study, the microwave data with optical data have been used to fulfill those temporal gaps. Data used were from RISAT-1 and Formosat-2 temporal data. The sensor details are shown as follows.

RISAT-1

RISAT-1 is Radar Imaging satellite 1, an Indian satellite that was built and operated by Indian space research organization (ISRO), carrying synthetic aperture radar (SAR), which operates on a C-band (5.35 GH_z) with dual polarization HH and HV in medium resolution scanSAR (MRS) mode. This satellite uses its active microwave remote sensing technique which can penetrate through the cloud and is capable of day–night imaging of the earth's surface. The purpose of this satellite was for applications in agriculture, particularly paddy monitoring in kharif season and management of natural disasters like floods and cyclones. The specifications of RISAT-1 are given in Table A2.2, and for Formosat-2 sensor details, refer to Table A2.1.

Methodology

A bi-sensor approach is used to analyze the paddy field with an effort to verify if SAR data can be incorporated with optical data with better classification results. The methodology followed for this work is shown in Figure A2.6.

Three datasets were generated from two different sensors, i.e., RISAT-1 and Formosat-2 with image acquisition on four different dates for the first case, i.e., for the transplant information of paddy. The dataset was created in such a way that it contained a minimum number of microwave data required to integrate with the optical data. For the dataset generation, temporal NDVI indices from the Formosat-2 images were calculated, and from the RISAT-1 data, backscatter coefficient property was calculated. These indices' information was taken at different dates, i.e., temporal NDVI indices as well as backscatter coefficient representing the information in the form of vector elements. Then elements of these vectors were taken into the classifier, and when this indices information club together it became the information of temporal indices. As it is shown in Figure A2.6, where vectors of temporal indices were generated, $\vec{ai}$ & $\vec{bj}$ represents backscatter coefficient vector and $\vec{ck}$ & $\vec{dn}$ represents NDVI vector information which were passed on together in the classifier. Tables A2.3 and A2.4 are shown the combination of three datasets.

TABLE A2.2
RISAT-1 Specifications

Band	C-band
Frequency	5.35 GHz
Polarization	Dual polarization (HH and HV)
Mode	Medium resolution scanSAR (MRS)
Revisit time	25 days

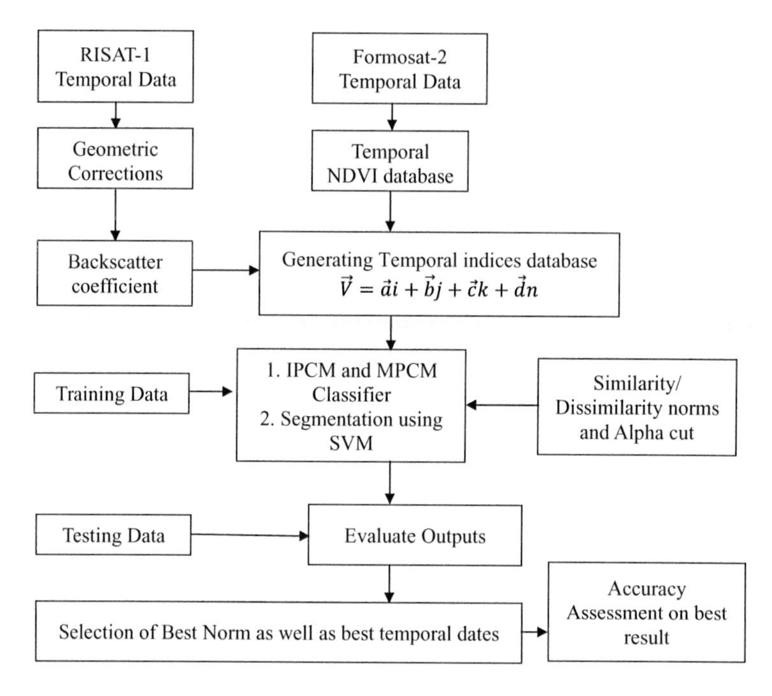

FIGURE A2.6 Methodology adopted for Case Study 2.

TABLE A2.3
Temporal Datasets for Late Transplant Paddy

	RISAT-1		Formosat-2	
	June 27, 2014	July 09, 2014	August 10, 2014	September 25, 2014
Dataset 1	✓	✓	✓	
Dataset 2		✓	✓	✓
Dataset 3	✓	✓	✓	✓

TABLE A2.4
Temporal Datasets for Harvesting Paddy

	FORMOSAT-2		RISAT-1	FORMOSAT-2
	September 25, 2014	October 10, 2014	October 31, 2014	November 03, 2014
Dataset 1	✓	✓	✓	
Dataset 2		✓	✓	✓
Dataset 3	✓	✓	✓	✓

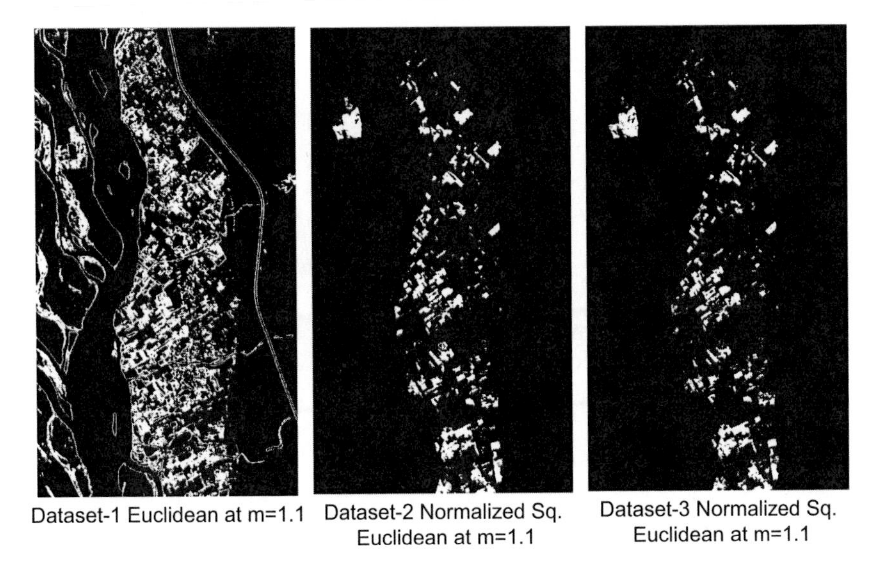

FIGURE A2.7 Improved possibilistic *c*-means results of late transplant paddy at optimized "m."

Results

Late transplanted paddy fields are successfully found with these classifiers, but for the harvested paddy fields, results are not encouraging. Hence, in the case of harvested paddy fields, image segmentation with SVM classifier is used. The very favorable results were obtained for late transplant paddy for IPCM classifier are normalized squared Euclidean norm at $m = 1.1$, while for MPCM classifier the best norms are normalized square Euclidean at $m = 1.1$. The best result is achieved for the late transplant paddy and harvested paddy while using the single temporal RISAT-1 image and two temporal Formosat-2 images (Figure A2.7). Overall accuracy of the late transplant paddy using IPCM and MPCM classifiers are 85.56%, and 81.11%, respectively. However, the accuracy of harvested paddy up to November 11 is found to be 72.22%.

CASE STUDY 3: HANDLING NON-LVINEARITY BETWEEN CLASSES USING KERNELS IN FUZZY CLASSIFIERS

In this study, kernel based fuzzy clustering has been used to handle both the problem of non-linearity and mixed pixels. A supervised kernel based fuzzy *c*-means classifier has been used to improve the performance of FCM classification technique. Eight kernel functions are incorporated to the objective function of the FCM classifier. As a result, the effects of different kernel functions can be visualized in generated fraction images. The best single kernel was selected by optimizing the weight constant which controls the degree of fuzziness using an entropy and mean membership difference calculation. Then top performing are combined to study the effect of composite kernels, which includes both the spatial and spectral properties. Fuzzy error matrix (FERM) was used to assess the accuracy of the results and was studied for AWiFS, LISS-III, and LISS-IV datasets from Resourcesat-1 and Resourcesat-2.

Objectives of the Case Study

The main objective of this research work is to optimally separate non-linear classes using a kernel based fuzzy *c*-means approach. The specific objectives were:

1. To develop an objective function for kernel based fuzzy *c*-means classifier (KFCM) (Chapter 5) to handle non-linear class separation.
2. To select the best single or composite kernel to be used within the KFCM classifier.
3. To evaluate the performance of this classifier in the case of untrained classes.
4. To study the best kernel model with the best possible parameter.

Study Area and Data Used

The study area considered for this particular research work was Sitarganj's Tehsil, Udham Singh Nagar district, Uttarakhand state, India. The considered area extends from 28°53′N to 28°56′N latitude and 79°34′E to 79°36′E longitude. Sitagarnj's Tehsil was recognized as it contained six land cover classes, e.g., agricultural fields with a crop, agricultural fields without crop both dry and moist, Sal and eucalyptus forests, and two water reservoirs: the Baigul (Sukhi) and Dhora reservoirs. The reasons for selecting this study area include:

1. Presence of mixed pixels which occurs because of degradation of land cover classes from one to another (water to grassland) will help to assess the capability of kernel based fuzzy *c*-means (KFCM) classifier.
2. Data from the sensors AWiFS, LISS-III, and LISS-IV from Resourcesat-1 and Resourcesat-2 were available from the same date to perform image-to-image accuracy assessment.

In this case study, AWiFS (Advanced Wide Field Sensor), LISS-III (Linear Imaging Self-Scanning System-III), and LISS-IV (Linear Imaging Self-Scanning System-IV) images of both Resourcesat-1 of IRS (Indian Remote Sensing Satellite) and Resourcesat-2 were used. Resourcesat-1 was launched in 2003, primarily for natural resource management with a 5–24 days repeat cycle. The images from AWiFS, LISS-III, and LISS-IV were acquired at the same time. The dataset available from Resourcesat-1 was captured at October 15, 2007 and from Resourcesat-2 at November 23, 2011 (Figure A2.8). The soft classified outputs from finer resolution LISS-IV images were used for the validation of the soft outputs of LISS-III and AWiFS. The specifications of the satellite data used are shown in Table A2.5.

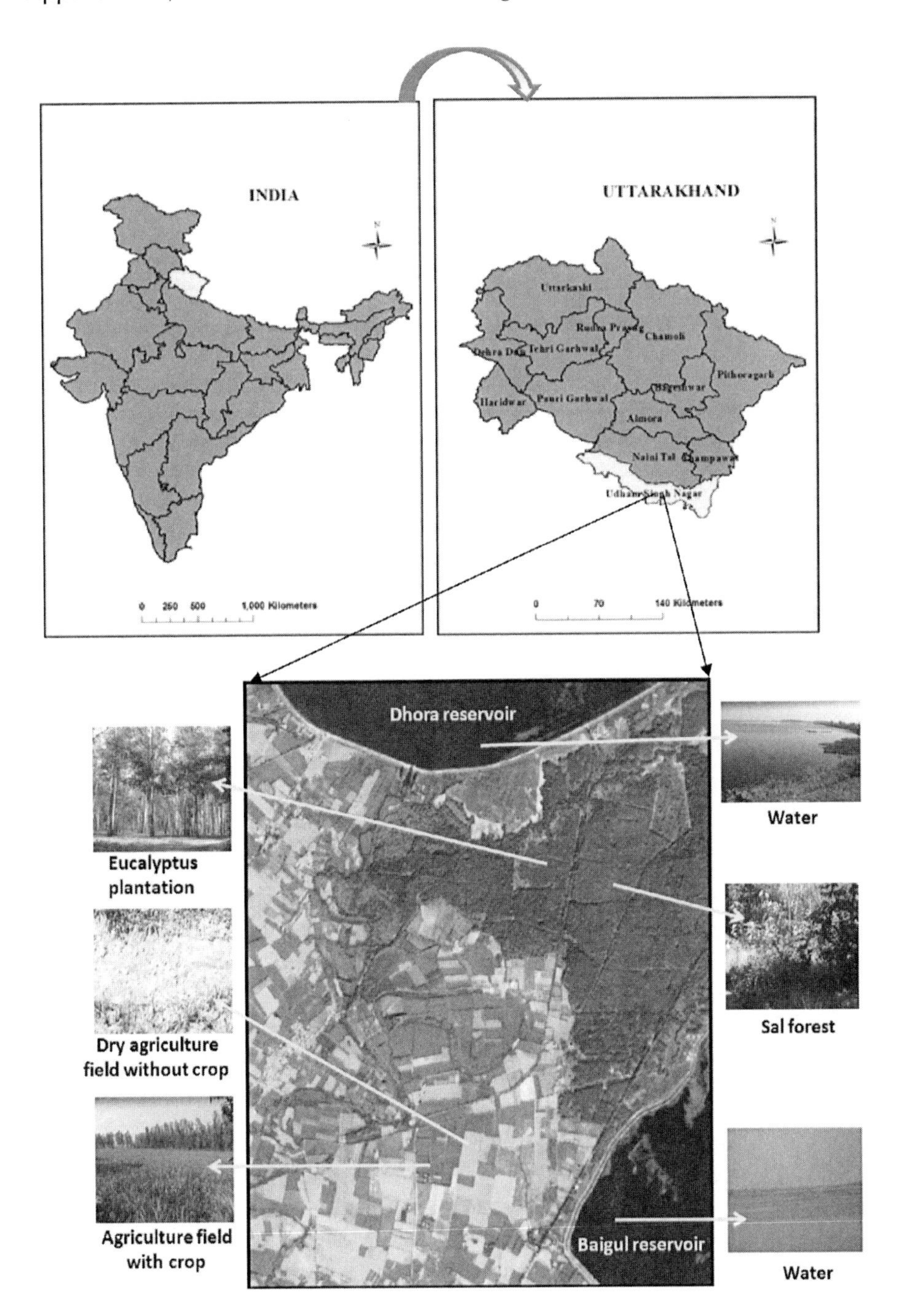

FIGURE A2.8 LISS IV (Resourcesat-2) image of Sitarganj's Tehsil with different land cover classes.

TABLE A2.5
Resourcesat-1 and Resourcesat-2 Sensors Specification

	AWiFS		LISS-III		LISS-IV	
Specification	**Resourcesat-1**	**Resourcesat-2**	**Resourcesat-1**	**Resourcesat-2**	**Resourcesat-1**	**Resourcesat-2**
Spatial resolution (m)	56	56	23.5	23.5	5.8	5.8
Radiometric resolution	10	12	7	10	7	10
Swath (km)	740	740	141	141	23.9 (Max Mode) 70.3 (Pan Mode)	70 (Max Mode) 70 (Mono Mode)
Spectral resolution (μm)	0.52–0.59 0.62–0.68 0.77–0.86 1.55–1.70	0.52–0.59 0.62–0.68 0.77–0.86 1.55–1.70	0.52–0.59 0.62–0.68 0.77–0.86 1.55–1.70	0.52–0.59 0.62–0.68 0.77–0.86 1.55–1.70	0.52–0.59 0.62–0.68 0.77–0.86	0.52–0.59 0.62–0.68 0.77–0.86

Methodology

The detailed explanation of steps adopted to achieve the objectives of this case study are shown by the flowchart in Figure A2.9.

Supervised KFCM classifier was adopted to generate the outputs of sub-pixel classification outputs. Three approaches, fuzzy *c*-means (FCM), FCM with single kernels (KFCM), and FCM with composite kernels, were considered for this study. Weight component m controls the degree of fuzziness, which was optimized based on the maximum mean membership difference between favorable and unfavorable classes and minimum entropy. Out of the three norms introduced in Chapter 3, Section 3.2.1, only one is considered, i.e., Euclidean norm. This is due to the fact that other norms, such as diagonal and Mahalonobis norms, are sensitive to noise and thus reduce the classification accuracy. This approach was adopted for a comparative study between simple FCM and the KFCM approaches.

Mainly three categories of kernels were considered: local kernels, global kernels, and spectral angle kernels. In this study, four local kernels were used: Gaussian kernel using Euclidean norm, radial basis kernel, kernel with moderate decreasing (KMOD), and inverse multi-quadratic kernel. Three global kernels were also used: linear kernel, polynomial kernel, and sigmoid kernel. Overall, eight single kernels were studied using the FCM approach. Followed by the implementation of eight single kernels, the next step was to optimize the weight component "m" using mean membership difference between favorable and unfavorable class method and entropy method. The best single kernels for each global and local category were selected based on the maximum mean membership difference between favorable and unfavorable classes and minimum entropy.

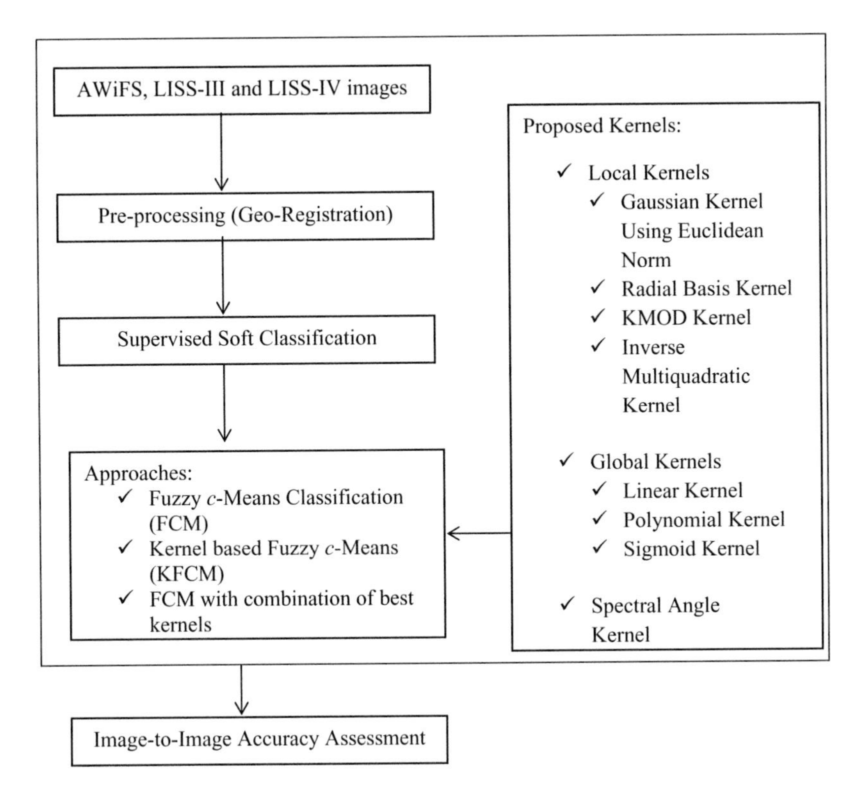

FIGURE A2.9 Methodology adopted for Case Study 3.

The composite kernels were obtained from the best single kernels. In composite kernels, the weight factor λ is given for each kernel which varies from 0.1 to 0.9. For composite kernels, the optimization of m and λ was necessary, and this was done considering maximum mean membership difference between favorable and unfavorable class and minimum entropy from where the best composite kernel was concluded. Untrained case outputs were also obtained by not training the KFCM classifier with the signature data of a class; in this study agricultural fields with crops under were considered as untrained class.

Results

Image-to-image accuracy assessment was conducted with reference datasets as LISS-IV for both AWiFS and LISS-III. For this, FERM was used to generate overall accuracy. The overall classification accuracy of KFCM classifier was compared with that of FCM classifier. Accuracy in the case of untrained case has been also evaluated. Overall, the highest fuzzy accuracy 97.03% was found from inverse multiquadratic kernel from Resourcesat-1 LISS-III dataset (Figure A2.10). Classification accuracy in the case of untrained classifier was also studied. A decrease in the average user's accuracy was observed when compared to trained cases.

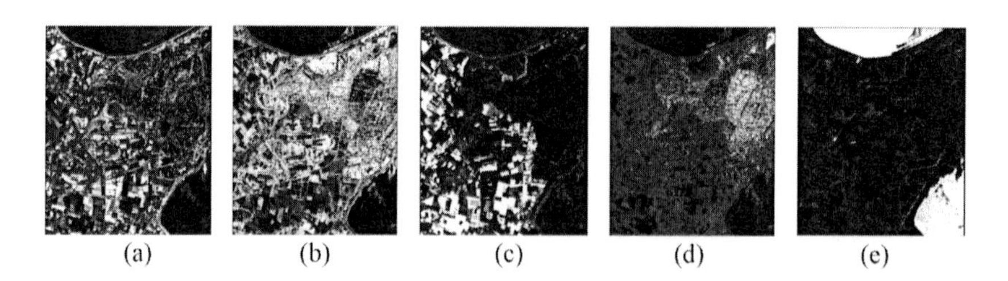

FIGURE A2.10 Generated fractional images for optimized values of *m* for Resourcesat-2 LISS-IV using inverse multi-quadratic for classes identified as (a) Agricultural field with crop (b) Eucalyptus plantation (c) Fallow land (d) Sal (e) Water.

CASE STUDY 4: HANDLING NOISE THROUGH MRF BASED NOISE CLUSTERING CLASSIFIER

Unlike conventional hard classifiers, fuzzy classifiers have been found to provide more realistic classification results. Independently, fuzzy classifiers succeeded in addressing the mixed pixel problem but not the isolated pixel problem, which was addressed by providing spatial contextual information to the classifier. Markov random field (MRF) has been identified as an effective technique to model spatial contextual information. This case study aims at realizing a hybrid fuzzy classifier by incorporating spatial contextual information into the noise classifier objective function. Spatial contextual information has been modeled using smoothness prior MRF (S-MRF) and four discontinuity adaptive MRF (DA-MRF) models, and their effect on the classification accuracy on coarser resolution dataset (e.g. AWiFS) and moderate resolution dataset (e.g. LISS-III) have been studied.

Objectives

The main objective of this research is to incorporate spatial contextual information with noise classifier using discontinuity adaptive MRF models. The specific objectives are:

1. To incorporate spatial contextual information with noise classifier (NC) using the smoothness prior model.
2. To study the effect of the four discontinuity adaptive MRF models when incorporated with the noise classifier.
3. To evaluate the performance of the noise clustering algorithm while applying discontinuity adaptive MRF models for both trained and untrained classes.

Study Area and Data Used (Refer to Case Study 3)

The geographical boundaries and the data used for this case study are identical to Case Study 3.

Methodology

In this study the main aim is to frame an objective function for noise classifier which incorporates the contextual information in an image using Markov random fields models, and includes smoothness prior MRF as well as discontinuity adaptive MRF models. Before incorporating the contextual information, the optimal parameters of NC need to be estimated so that the noise classifier could be tuned to its best performance.

Before classification, the LISS-III, LISS-IV, and AWiFS images from Resourcesat-1 and Resourcesat-2 sensors have been geometrically corrected and geo-registered. A survey of India toposheet has been used initially to geometrically correct the LISS-IV dataset which was then used for the geo-registration of AWiFS and LISS-III datasets. This would be followed by a classification of data by NC, NCS-MRF, and four different NC DA-MRF classifiers. Finally, the accuracy will be accessed using soft classified finer resolution data from LISS-IV sensor and also with the help of field observation data. The performance of the classifier will be evaluated in case of untrained classes as noise/outliers also (Figure A2.11).

Results

FERM has been used in this case study for accuracy assessment and the reference data was generated from the high resolution LISS-IV datasets. Most DA-MRF models improved the classification accuracy. NC DA4-MRF has been found to be the most accurate, at the same time, having the best edge preserving capability among other MRF models (Figure A2.12). The accuracies of classification for AWiFS, LISS-III datasets from Resourcesat-1 and Resourcesat-2 have been found to be 87.26%, 89.40%, and 85.27%, 89.37%, respectively. For the untrained case, a small decrease in accuracy has been observed.

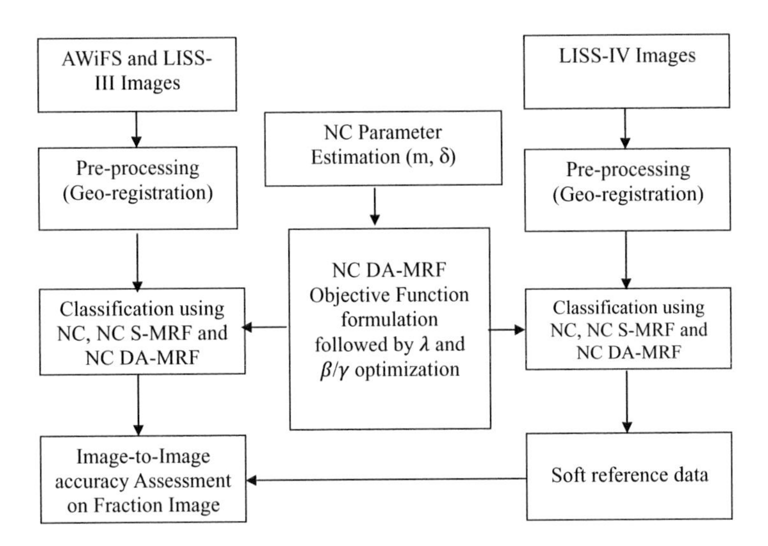

FIGURE A2.11 Methodology adopted for Case Study 4.

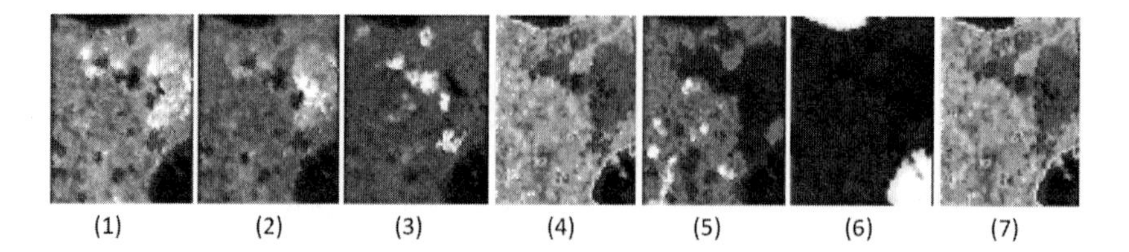

FIGURE A2.12 Fractional images obtained from NCMRF-DA4 classifiers on AWiFS dataset from Resourcesat-1. The fractional images correspond to Agriculture fields with crop (1), Sal Forest (2), eucalyptus plantation (3), dry agricultural field without crop (4), moist agricultural field without crop (5), water (6), noise (7).

CASE STUDY 5: LOCAL CONVOLUTION BASED CONTEXTUAL INFORMATION IN POSSIBILISTIC *c*-MEANS CLASSIFICATION

In this study, three FCM based spatial contextual classifiers, FCM-S, FLICM, and ADFLICM (Chapter 5), were selected and their performance has been examined on remotely sensed images. These classifiers use spatial information within a defined local neighborhood of each pixel for estimating the class membership of that pixel. The FCM-S, FLICM, and ADFLICM classifiers incorporate spatial information by using a simple approach of modifying the FCM objective function to include a term that controls the effect from neighboring pixels. Further, this case study approach has been used to modify the PCM classifier to develop PCM based local spatial information classification algorithms such as PCM-S, PLICM, and ADPLICM.

Objectives of the Case Study

The main aim of this case study is to compare the ability of fuzzy local spatial information classifiers in handling ambiguities, caused by within-class spectral variation and spectral similarity among different classes, without losing essential image details. The specific aims include:

1. To study and compare the performance of three existing FCM based on local spatial information classification algorithms for land cover classification.
2. To apply fuzzy local spatial information algorithms with PCM as the base classifier and analyze their performance in noise reduction (isolated pixels) and preservation of image details.
3. To conduct a comprehensive assessment of the PCM based local spatial information classifiers and the FCM based local spatial information classifiers.
4. To examine the effects of some key parameters such as fuzzifier, window size, and the factor that controls the impact of neighborhood on the performance of these algorithms and optimize these values.

Study Area and Data Used

The geographical location of the study area is identical to the Case Study 1. Remote sensing data from Landsat-8 sensor at 30 m spatial resolution and Formosat-2 sensor at 8 m spatial resolution have been used in this study. The classification algorithms have been executed on the Landsat-8 image, and the Formosat-2 image has been used to create a reference map for accuracy assessment. All the bands of the images were used for classification and reference map generation. Figure A2.13 shows the satellite images of the study area with the six LULC classes. The specifications of both the datasets are summarized in Table A2.6.

Methodology

Figure A2.14 explains the steps followed for the development, execution, and assessment of the performance of these algorithms.

Results

Supervised classification with the algorithms developed was performed on Landsat-8 image. For the validation of results, soft reference data was created from a finer resolution Formosat-2 image of the same study area captured around the same time.

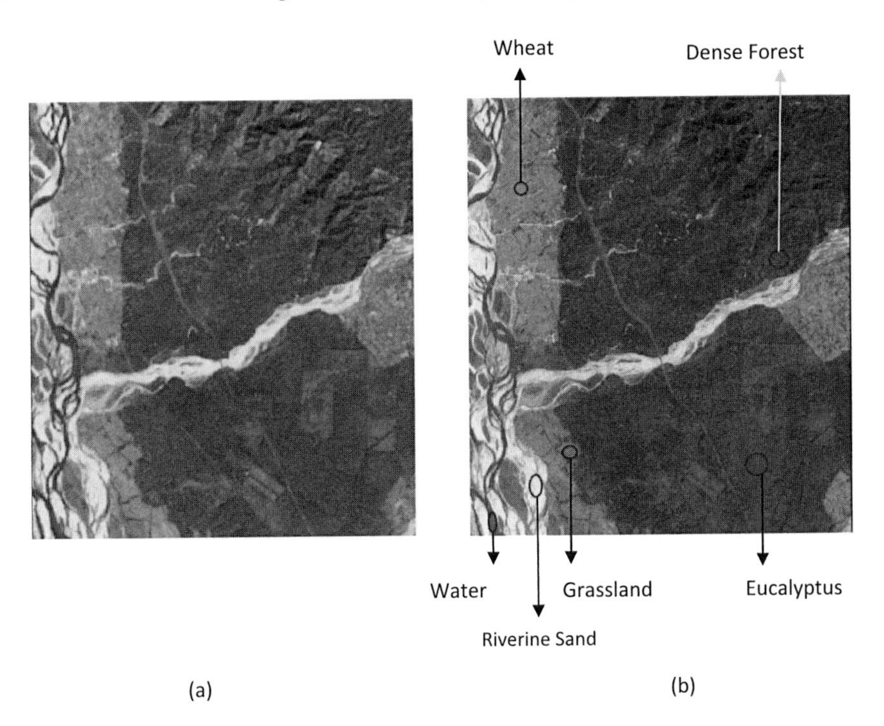

FIGURE A2.13 The datasets with identified LULC classes. (a) Landsat-8 image used for classification. (b) Formosat-2 image with identified LULC classes in the study area.

TABLE A2.6
Specifications of the Datasets Used

Specification	Landsat-8	Formosat-2
Spatial resolution (multi-spectral)	30	8 m
Spectral resolution	8 Bands	4 Bands
	Band 1-Coastal	Band 1-Blue
	Band 2-Blue	Band 2-Green
	Band 3-Green	Band 3-Red
	Band 4-Red	Band 4-NIR
	Band 5-NIR	P is Panchromatic (2 m)
	Band 6-SWIR1	
	Band 7-SWIR2	
	Band 9-Cirrus	
	Band 8 is Panchromatic (15 m)	
Sensor footprint	170 × 185 km	24 × 24 km
Revisit interval	16 Days	Daily
Date on which image was acquired	February 12, 2015	February 21, 2015

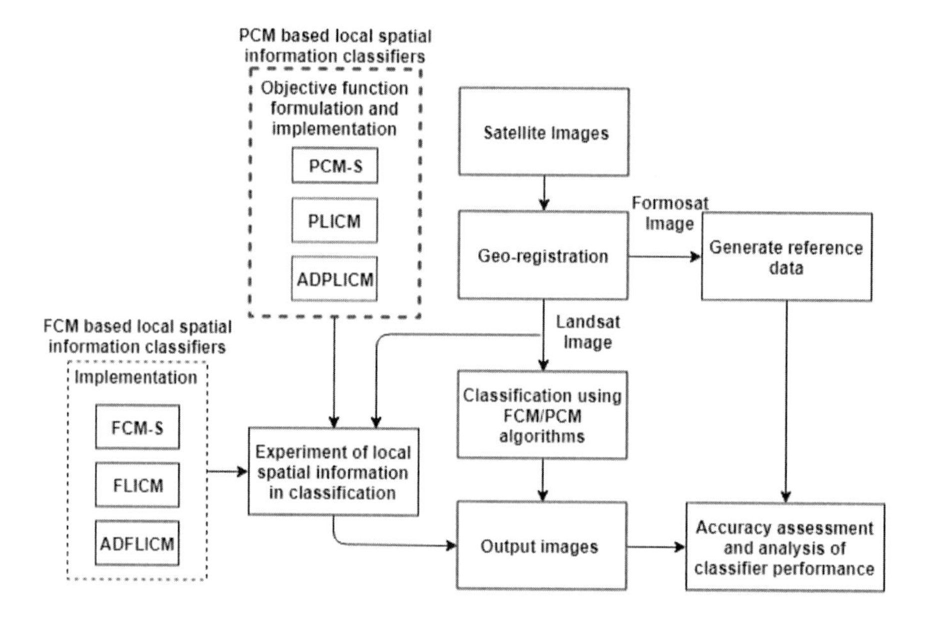

FIGURE A2.14 The methodology adopted for Case Study 5.

Various experiments were conducted to analyze the performance of the FCM and PCM based local spatial information classification algorithms developed. To better analyze the effect of incorporating spatial information into base classifiers, the performance of the classifiers was compared to the respective base classifiers (FCM and PCM).

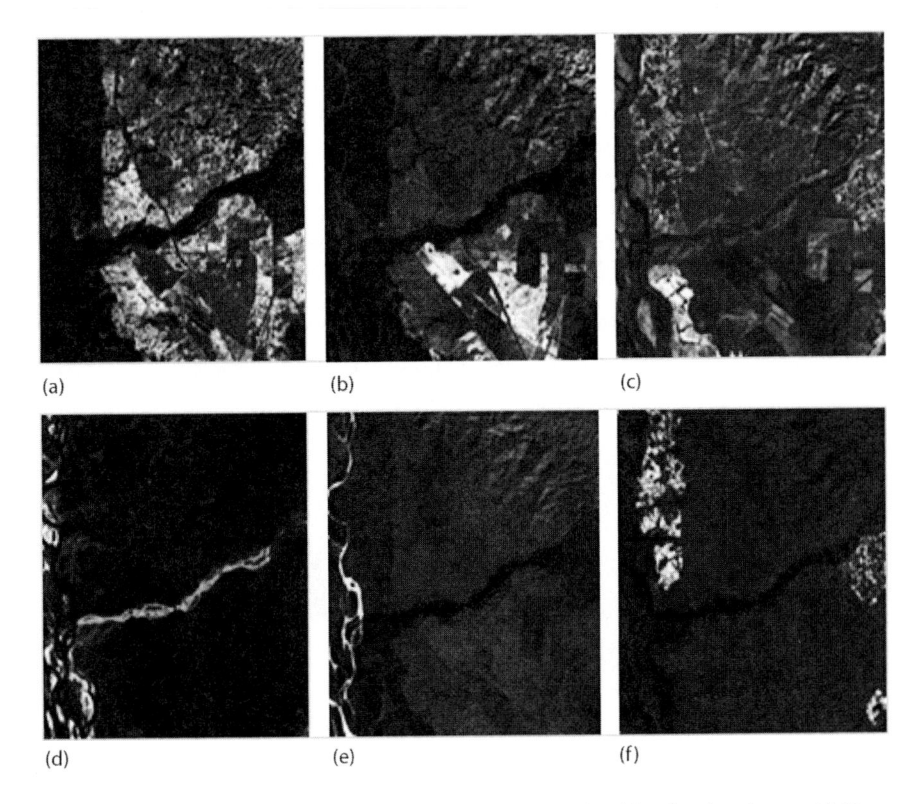

FIGURE A2.15 Output fraction images of ADPLICM classifier for the classes (a) Dense forest (b) Eucalyptus (c) Grassland (d) Riverine sand (e) Water (f) Wheat.

Fuzzy error matrix (FERM), root mean square error (RMSE), and mean membership difference methods were used to analyze the performance of the classification algorithms quantitatively. The results suggested that FCM based and PCM based local spatial information classifiers outperform the conventional FCM and PCM based classifiers in terms of overall accuracy (Figure A2.15). However, the local spatial information algorithms may produce over smooth results, which may cause loss of image details. It can be inferred that an appropriate local spatial information is necessary for optimal performance.

CASE STUDY 6: OPTIMIZATION OF LOCAL CONVOLUTION BASED MPCM CLASSIFIER AND IDENTIFICATION OF PADDY AND BURNT PADDY FIELDS

In this case study, the local information methods have been deployed with possibilistic fuzzy based classifiers by adding spatial constraints, fuzzy factor, local similarity measure/pixel spatial attraction model, and MRF model in order to preserve image details and aiming to provide robustness to noise and outliers. Local information correlates the information between a pixel and its adjacent pixels in the image space. This research work presents PCM-S, PLICM, ADPLICM, MPCM-S, MPLICM, ADMPLICM algorithms (Chapter 5) in a supervised classification. These algorithms can overcome the limitations of PCM, MPCM, PCM-MRF, and MPCM-MRF algorithms by incorporating local information through local convolution. PCM-S and MPCM-S have been deployed by adding spatial constraints through local convolution (N_R) and (α) to supervise the effect of the neighbors term in conventional PCM and MPCM algorithms, respectively; PLICM and MPLICM have been deployed by adding fuzzy factor (G) in conventional PCM and MPCM algorithms, respectively, to preserve more image details; ADPLICM and ADMPLICM have been deployed by adding local similarity measures and pixel spatial attraction model (S_{ir}) to sustain image details and aiming to provide robustness to noise in conventional PCM and MPCM algorithms, respectively.

Objectives

The overall objective of this case study was to include local information in fuzzy based classifier. The specific objectives to meet the desired requirements are:

1. To incorporate local information through convolution.
2. To incorporate local information through MRF.
3. To study the effect of various local information methods.
4. To study the effect of convolution window size.
5. To test the approach for burned areas in harvested paddy fields.

Study Areas and Data Used

Two study areas were chosen for this research. First was from Haridwar site as used in Case Study 1, while another study area has been considered for identification of burnt-paddy field located in Patiala District, Punjab State, India. The stubble burning phenomenon in Punjab is linked to three unlikely factors: the (relatively) large size of landholdings of farmers in state, the (consequent) high level of mechanization, and water conservation process that shortens the harvest window. In terms of geographic coordinates, the area ranges from 30°26′22.8657″N, 76°5′30.4380″E to 30°10′0.4143″N, 76°23′58.1095″E. The area consists of agricultural farm paddies as one of the major crops, and has some eucalyptus plantations. Figure A2.16 shows the geographic location of the study area, and the reasons for selecting the study area is that the stubble:

Stubble burning is present during October (2018)–November (2018) in this region. The harvesters in Punjab use to shave-off the grainy part of paddy, leaving loose straw in their wake, with an observation that it is cheaper to clear the residue by burning.

In this research work, Landsat-8, Formosat-2, and Sentinel-2A/2B temporal images were used. The Landsat-8 image was used for classification, Formosat-2 for reference data, and temporal images of Sentinel-2A/2B were used for further testing of proposed algorithms. The dataset used in this research with sensor specifications are mentioned in Table A2.7.

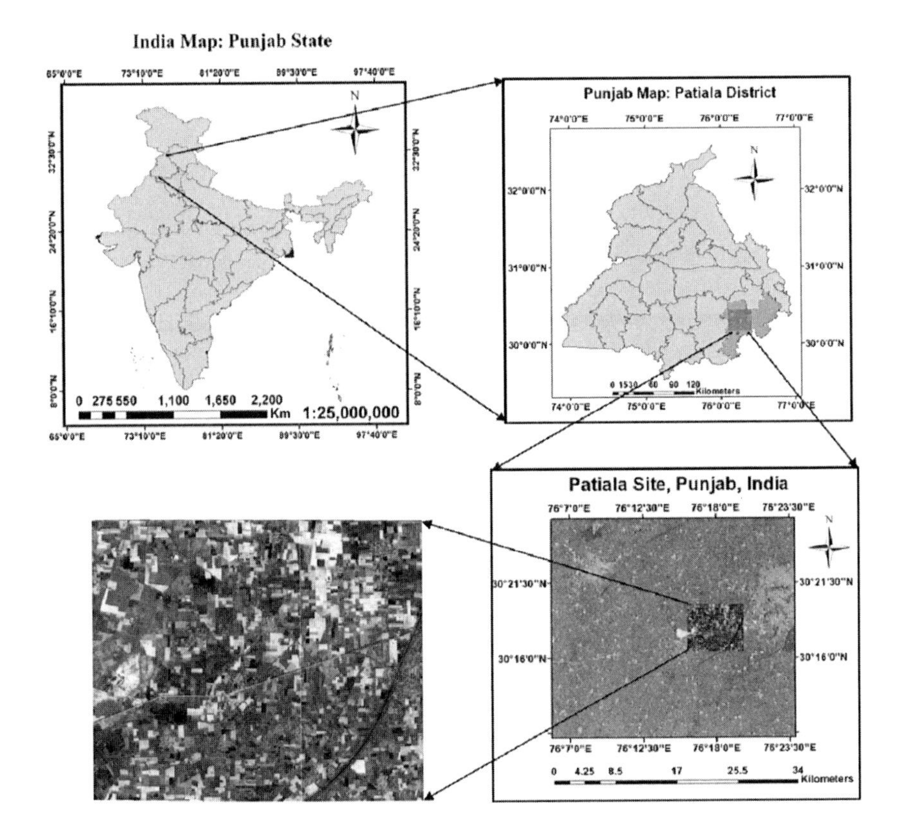

FIGURE A2.16 Study area-2 (Patiala Area, Punjab, India).

In order to make a proper identification of stubble burnt patches, multi-spectral and temporal Sentinel-2A/2B data (10 m resolution) of September 30, October 5, October 12, October 15, October 20, October 25, October 30, November 9, and November 19, 2018 were selected. Blue, green, red, NIR, SWIR-1, and SWIR-2 has been taken for this study. Figure A2.17 shows the burnt-paddy field sites that has been identified on October 27, 2018, and October 28, 2018, in Patiala District.

Methodology

The proposed methodology adopted for this research has been mentioned in Figure A2.18. The complete methodology for this study can be divided into five stages:

1. *Data Pre-processing:* Steps like atmospheric correction and geo-registration were done.
2. *Development of Algorithm:* Local Information was incorporated in PCM, MPCM using PCM-S, PLICM, ADPLICM, MPCM-S, MPLICM, ADMPLICM, MRF-SP, and MRF-DA algorithms.

TABLE A2.7
Landsat-8, Sentinel-2, and Formosat-2 Sensor Specifications

	For Local Convolution Study (Haridwar Area)		Burnt-Paddy Field Identification (Patiala Area)
Details	**Reference Data**	**For Classification**	
Specifications	**Formosat-2**	**Landsat-8**	**Sentinel-2A & 2B**
Spatial resolution	8 m for MS imagery	30 m for MS imagery	10 m (B2,B3,B4,B8), 20 m (B11,B12)
Spectral bands	B1 = 450–520 nm (Blue) B2 = 520–600 nm (Green) B3 = 630–690 nm (Red) B4 = 760–900 nm (NIR)	B1 = 430–450 nm (Coastal) B2 = 450–510 nm (Blue) B3 = 530–590 nm (Green) B4 = 630–670 nm (Red) B5 = 850–880 nm (NIR) B6 = 1570–1650 nm (SWIR 1) B7 = 2110–2290 nm (SWIR 2) B9 = 1360–1380 nm (Cirrus)	B2-Blue (490 nm) B3-Green (580 nm) B4-Red (665 nm) B8-NIR (842 nm) B11-SWIR-1 (1610 nm) B12-SWIR-2 (2190 nm) **Temporal Multi-spectral data** Sep 30, 2018; Oct 05, 2018; Oct 12, 2018; Oct 15, 2018; Oct 20, 2018; Oct 25, 2018; Oct 30, 2018; Oct 09, 2018; Oct 19, 2018;
Swath width	24 km	185 km	290 km
Pixel quantization	12 bit	16 bit	12 bit

FIGURE A2.17 Burnt-paddy field sites identified on October 27, 2018, and October 28, 2018, in Patiala district.

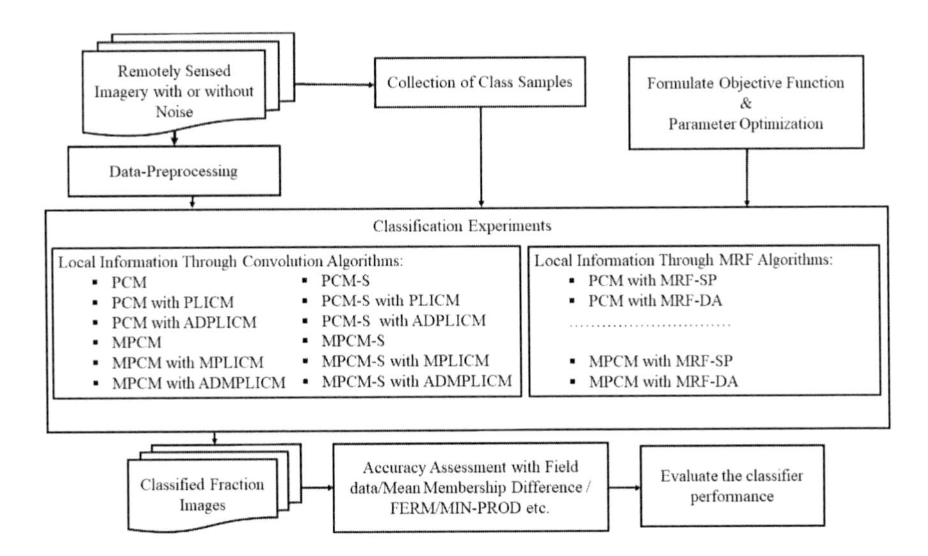

FIGURE A2.18 Case study methodology for optimizing parameters.

3. *Parameter Optimization:* Parameter optimization has been conducted for all algorithms using independent MMD technique.
4. *Classification:* Classification of FORMOSAT-2 & LANDSAT-8 images have been performed using proposed algorithm.
5. *Accuracy Measure:* Accuracy assessment of classified fraction images w.r.t. classified reference image using root mean square error (RMSE), fuzzy error matrix (FERM), and MIN-PROD methods.

Region growing for collection of training data: This is basically a segmentation technique for identifying homogeneous behavior of pixels in an image. This technique has several advantages over conventional Euclidean measure by incorporating spectral characterization measures that sets apart pixels on the basis of their spectra analyzed using divergence and correlation function of measures. Benefits of region growing technique are as follows:

1. Maximum number of training sample collection
2. Provision of incorporating spectral measures.

Methodology adopted for burnt-paddy field identification: The objective of using this test site was to test the performance of modified possibilistic based algorithm on different burnt sites of Patiala District. Temporal multi-spectral data of Sentinel-2A/B from September 30, 2018, to November 14, 2018, have been used for burnt-paddy field identification. The process of implementation follows several steps:

1. Resampling of Band-11, Band-12 from 20 m spatial resolution to 10 m spatial resolution then stacked with Band-2, Band-3, Band-4, and Band-8.
2. The next step involves the collection of training samples using region growing technique and then calculation of indices that have been fed to the classifier.
3. Formulate objective function and optimized required parameters for selected possibilistic based algorithm for classification.
4. Classified results have been verified with field data and NBR (normalized burnt-ratio) and dNBR (delta normalized burnt-ratio).
5. Extraction of burnt-paddy field has been identified at different sites with five to six day intervals and the performance of classifiers has been evaluated.

The flow diagram for this study is shown in Figure A2.19. While identifying paddy stubble burnt field's using temporal remote sensing data, the MPCM classification algorithm has been applied. Modified possibilistic fuzzy based classifier is capable to extract only single class of interest like paddy stubble burnt field for each 5 day interval while reducing spectral dimensionality of temporal remote sensing data, using class based sensor independent indices approach.

Results

Three experiments have been conducted to optimize, test, and validate the proposed algorithms, i.e., mean membership difference (MMD), root mean square error (RMSE), and fuzzy error matrix (FERM)/MIN-PROD, respectively. MMD has been calculated on original classified images for independent accuracy to optimize

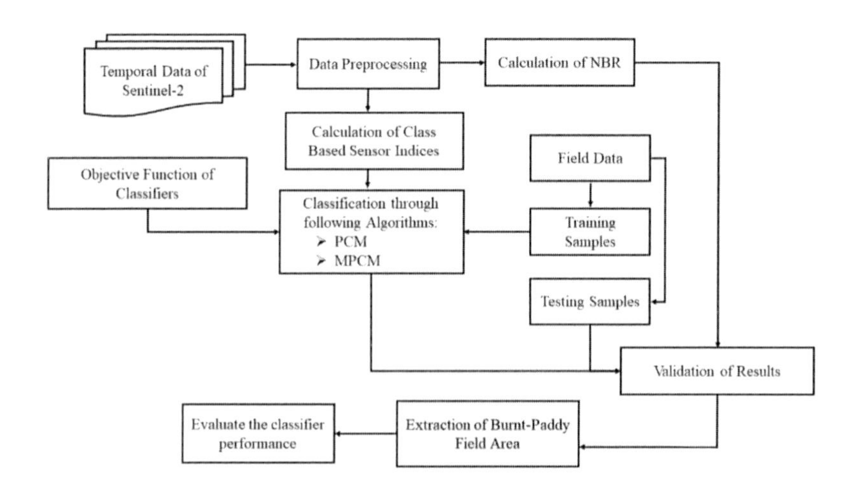

FIGURE A2.19 Flow diagram adopted for burnt-paddy field identification.

the parameters of proposed algorithms for Landsat-8 and Formosat-2 satellite imagery. RMSE has been calculated between original and noisy classified images for Formosat-2 satellite imagery in which MPCM-S produces least value. FERM and MIN-PROD composite operators have been used with respect to soft reference data. The result depicts that proposed PCM and MPCM classifier with local information can improve the fuzzy overall accuracy and fuzzy kappa by incorporating local information for remotely sensed imagery. Therefore, it was found that PCM with ADPLICM with Euclidean norm in possibilistic based fuzzy classifier and MPCM-S with ADPLICM classifier with Euclidean norm in modified possibilistic fuzzy classifier performed better as compared to other proposed algorithms.

The MPCM-S algorithm after training and optimization for temporal multispectral Sentinel-2A/B data was used for identification of burnt-paddy field using temporal class based sensor independent indices database. This data was analyzed and detected in temporal dataset of Sentinel 2A/B from September 30, 2018, to November 19, 2018. The result showed that stubble burning was started in a more frequent way from October 20, 2018, to November 19, 2018 (Figure A2.20).

FIGURE A2.20 Burnt patches on November 9, 2018.

CASE STUDY 7: SEMI-SUPERVISED TRAINING APPROACH FOR PCM CLASSIFIER

The demand for good classification accuracies with minimized efforts has led to options where a few labeled data could enhance the classification accuracies. In this case study, a semi-supervised approach is considered to minimize the efforts in collecting huge labeled training samples, which is often strenuous and time-consuming. To handle uncertainties better, PCM classifier is used and considered with different approaches as per the labeled training data. The similarity between two pixel vectors becomes important when it comes to collecting unlabeled data in a semi-supervised approach, or when the classification is performed based on assigning more membership values to similar pixels. The hybrid spectral similarity measures, spectral information divergence with spectral angle measure and spectral information divergence with spectral correlation angle, are used to measure the similarity between two pixel vectors due to their proven capabilities for capturing high band-to-band variability for hyperspectral imagery. Their roles are studied in view of a multi-spectral imagery. The PCM classifier is used with semi-supervised training data using different hybrid measures. Due to the availability of a few labeled training samples, the mean shift algorithm is employed to refine the training data and to shift the mean for a PCM classification algorithm to a higher density region. In addition, the proposed methods relate the bandwidth parameter from the mean shift algorithm to the bandwidth or resolution parameter of PCM classifier with an iterative procedure to capture the class variances. The methods are applied to input LANDSAT-8 imagery with 30 m spatial resolution. Formosat-2 images have been used for reference data to test the performance using the FERM.

OBJECTIVES OF THE CASE STUDY

The main objective of this study was to find out the role of hybrid spectral similarity measures for a semi-supervised possibilistic fuzzy classifier in classifying multi-spectral imagery. The sub-objectives proposed to reach the main objectives are as follows:

1. To study the effectiveness of the proposed hybrid spectral similarity measures for multispectral imagery.
2. To develop a precise signature by using hybrid spectral similarity measures for a multispectral image.
3. To identify and apply a suitable approach for incorporating a semi-supervised learning method in a possibilistic fuzzy classifier.
4. To develop and optimize an objective function for a semi-supervised possibilistic fuzzy classifier with hybrid spectral similarity measures.
5. To compare the performance of hybrid spectral similarity measures with conventional similarity measures (Euclidean distance).
6. To compare the performance of a semi-supervised approach with the supervised approach.

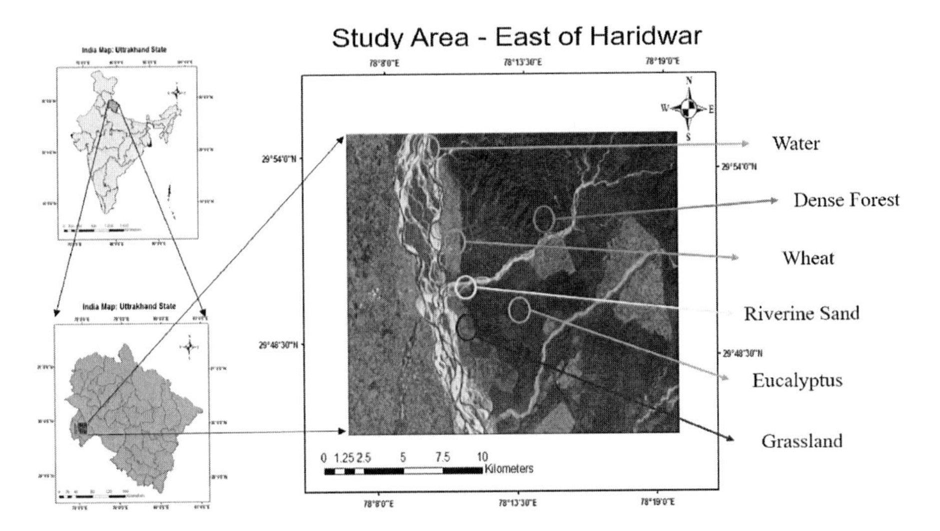

FIGURE A2.21 Location of major land cover for the study area.

Study Area and Data Used

The geographical location of the study area is identical to Case Study 1(Figure A2.21). The acquisition and other details of satellite data are identical to Case Study 5.

Methodology

Case 1: When membership values are initialized first.

Initial membership values: Obtained from labeled training data

η: Depends only on labeled training data (initial memberships)

The steps are described as follows:

- STEP 1: Initialize the class memberships from the labeled training data.
- STEP 2: Calculate mean and distance of each pixel to this mean.
- STEP 3: Calculate η using the initialized class memberships.
- STEP 4: Calculate final membership values of every pixel to every class.

Case 2: When mean value is estimated first.

Initial membership values: Obtained from all pixels

η: Depends on all pixel values (estimated initial memberships)

The approach has been shown in Figure A2.22, and the steps are described as follows:

- STEP 1: Initialize the class mean values from the labeled training data.
- STEP 2: Calculate the distance of each pixel to this mean.
- STEP 3: Calculate initial membership values using the calculated distance.
- STEP 4: Calculate η using the initialized class memberships.
- STEP 5: Calculate final membership values of every pixel to every class.

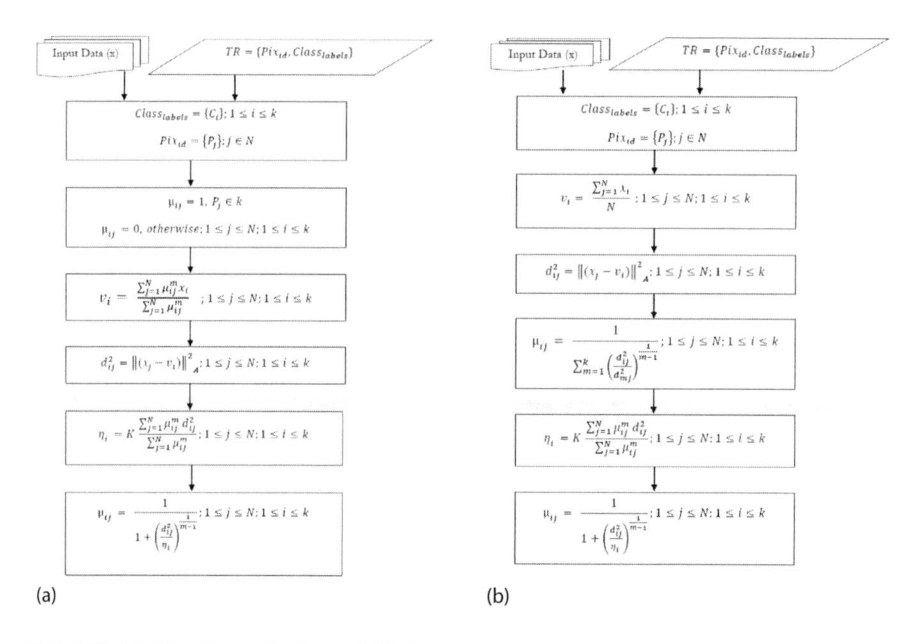

FIGURE A2.22 Supervised possibilistic *c*-means approach: (a) Case 1 and (b) Case 2. *Note:* The major difference in both the approaches lies in the estimation of initial class membership values μ and η. Both the approaches tend to give similar estimates in case of large and reliable training data (supervised approach).

Semi-supervised PCM Classifier

The semi-supervised PCM clustering algorithm uses a very few labeled training samples and a large number of unlabeled samples for the calculation of parameters such as mean, memberships, and η. The following diagram as represented in Figure A2.23 depicts the generalized semi-supervised approach to classification with a possibilistic *c*-means classifier.

The semi-supervised possibilistic *c*-means classifier can be both labeled and unlabeled. Therefore reliable estimates of η and class membership values μ could not be obtained due to the following reasons:

- The labeled data is very small in number as compared to the unlabeled data.
- The unlabeled data obtained and added may or may not be a complete representation of a class.
- The parameters such as mean, η, and class membership values μ obtained from the semi-supervised signature data may or may not represent entire variance for the class.

Results

The proposed methods are compared with the conventional methods such as the supervised approach and Euclidean distance as a similarity measure. It is found that the results from a completely supervised approach are comparable with the results

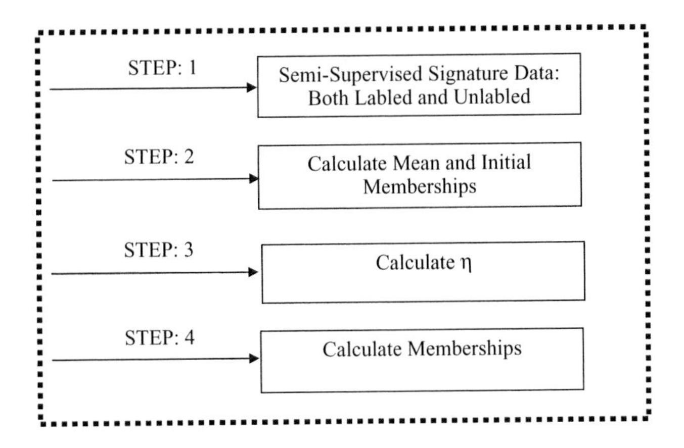

FIGURE A2.23 Generalized procedure for semi-supervised possibilistic *c*-means.

of a semi-supervised approach when the PCM classifier is incorporated with the respective distance measures. In comparison to hybrid measures, Euclidean distance is found to be the best in terms of capturing the high inter-class and intra-class variability when incorporated as a distance measure in the PCM classifier and also in measuring the similarity between two pixels with highest overall accuracies and lowest global root mean square error. Among the hybrid measures, spectral information divergence with spectral correlation angle works best in terms of measuring the similarity between pixel vectors and as a distance measure in a PCM classifier. Also, the hybrid measure spectral information divergence with spectral angle measure works best for Eucalyptus class where intra-class variability is low (Figure A2.24).

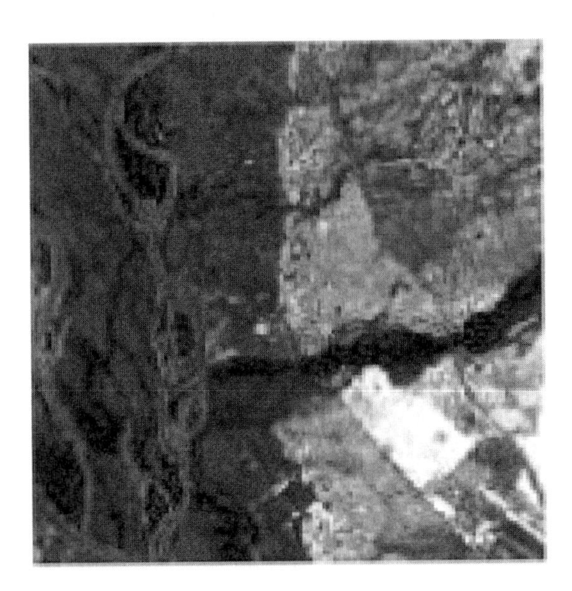

FIGURE A2.24 Output from semi-supervised approach.

In addition, the classification methods with just a few labeled samples, after shifting the mean to a higher density region using the mean shift algorithm, give comparable results to both supervised (large labeled data) as well as semi-supervised approaches, with respective distance measures. Also, it is possible to capture variance and achieve higher classification accuracies with a few labeled training samples using the mean shift algorithm and a PCM classifier.

CASE STUDY 8: STUDY OF HYBRIDIZING STOCHASTIC AND DETERMINISTIC MEASURES WITH FUZZY BASED CLASSIFIER

Spectral measures have been tested and proven to deliver satisfactory results for hyperspectral images. This case study aims to test their applicability and potential on multispectral images using both stochastic and deterministic spectral measures. The concept of hybridizing stochastic and deterministic measures on a soft computing platform has been introduced for assessing performance of hybrid measures to achieve the objectives of the study. To further enhance the potential of selected spectral information divergence (SID), spectral angle mapper (SAM), and spectral correlation angle (SCA) measures, we utilize a region growing algorithm for collection of accurate and precise training data.

Objectives of the Case Study

The main objective of this research is to study the performance of spectral characterization measures based FCM. The main objective can be split into the following sub-objectives which can be stated as:

1. To study performance of single proposed measures SID, SAM-TAN, SAM-SIN, SCA, SCA-TAN, and SCA-SIN on multi-spectral data.
2. To study performance of hybrid proposed measures SIDSAM-TAN, SIDSAM-SIN, SIDSCA-TAN, and SIDSCA-SIN on multi-spectral data.
3. To make a comparative study of single and hybrid proposed measures with conventionally used Euclidean measure.
4. To study the effect of the region growing algorithm used to collect training samples on classification accuracy.

Study Area and Data Used

The geographical location of the study area is identical to Case Study 1. The acquisition and other details of satellite data are identical to Case Study 5. A simulated image of Formosat-2 as in Case Study 1 (Figure A2.2) has been used to study the behavior of developed KPCM algorithm accurately. The advantages of using simulated images are as follows:

- The composition of each class is known.
- The classes can be mixed in different proportions.

- The capability to handle the mixed pixel by the developed KPCM algorithm can also be verified.
- The pixels within the class can be easily located within the simulated image.

Methodology

The methodology adopted to fulfill the objectives of the study utilizes two multi-spectral images, Formosat-2 of 8 m spatial resolution and Landsat 8 of 30 m resolution. When FCM is implemented on an image, it becomes necessary to determine the value of "*m*" for which an optimum classification is performed (Chapter 3, Equation 3.1). Therefore, the initial step to implement the FCM algorithm with spectral characterization measures begins with optimization of measures which determines best suited value of "*m*" for optimum classification.

The optimization of measures is performed using a simulated image generated using class signatures collected from satellite images. A simulated image of 900 rows and 900 columns is created with three types of zones. A simulated image consists of a single class zone, a 2-class mixing zone with a 50:50 ratio, and a 3-class mixing zone with a ratio 30:40:30. There is an intentional addition of interclass variation of 1 unit added for each single class.

This image is utilized to optimize all measures by obtaining membership values from all three zones by gradually varying the value of weighted constant "*m*." The membership values for each zone are acquired after classification of simulated image form ϵ [1, 3] at an interval of 0.1. This process is repeated for each spectral measure. Thereafter a mixed pixel class variance analysis is carried out for each measure based on membership values acquired from classified simulated image. In a 2-class mixing region, we have a class mixing of ratio 50:50. As in an image, the maximum membership of a pure pixel is 1. Thus, each class in a 2-class mixing region gets 0.5 class membership. So, to calculate Mixed Pixel Class Variance (MPCV), the obtained membership value from class-1 and class-2 of the two class mixing region is deducted by 0.5, whose absolute value is accepted as MPCV for that particular value of "*m*." Similarly, for a 3-class mixing region where classes are mixed in a ratio of 30:40:30, the class with a share of 40% gets membership value deducted by 0.4 and classes with a share of 30% get their membership value deducted by0.3, and the absolute value of their sum serves as MPCV of the 3-class mixing region.

Subsequently, the spectral measures get optimized and are ready to operate on both satellite images, Formosat-2 and Landsat-8 multi-spectral images. The implementation of fuzzy *c*-means objective function on multi-spectral images is based on three major aspects:

1. Training data collected for each class acquired by ground survey from the image.
2. Optimization values obtained from simulated image by MPCV analysis.
3. Selection of spectral characterization measure that drives FCM algorithm.

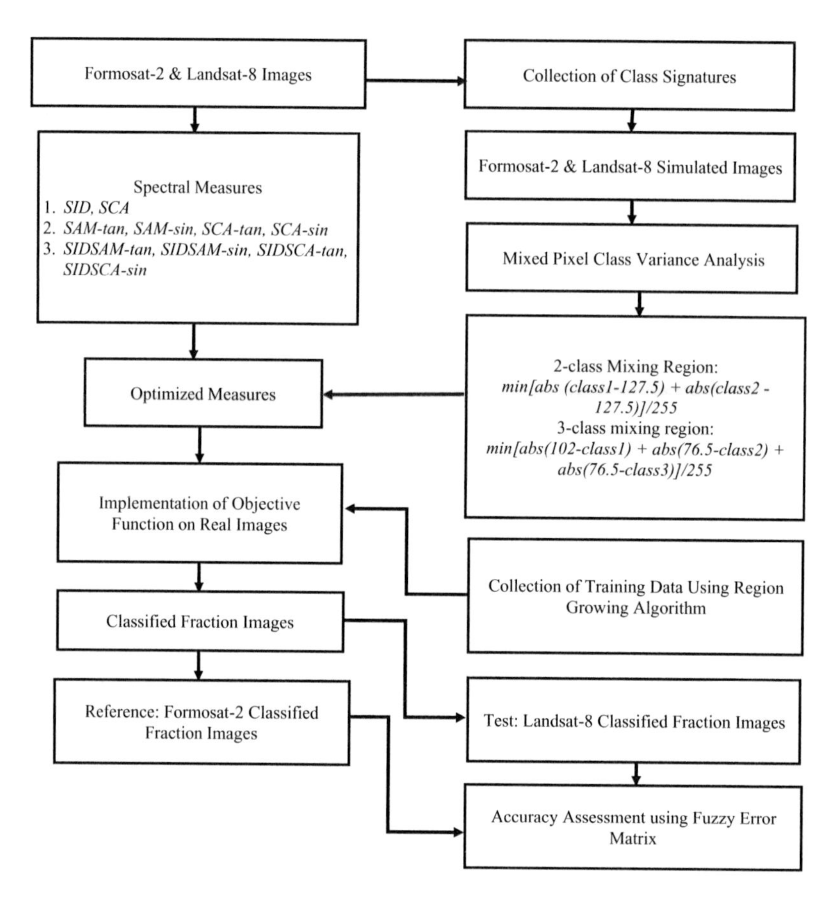

FIGURE A2.25 Methodology observed to carry out Case Study 8.

The objective function was implemented on both Formosat-2 and Landsat-8 multispectral images. The FCM algorithm works to allocate membership value to each pixel based on training data provided for all classes. Consequently, the fuzzy *c*-means algorithm generates a fraction image of each class. The detailed flowchart of methodology observed to carry out the study is shown in Figure A2.25.

Results

As expected, a large numbers of collected training samples enable spectral measures to perform more accurately. Experiments reveal superiority of stochastic measure (SID) of achieving highest classification accuracy of 72.86% calculated using FERM. However, it was observed that if a soft computing algorithm is trained with numerous region grown accurate training samples, the accuracy of measures is subjected to an increment of 5–8%. As a result, stochastic SID measure ended up giving 80.4% of overall accuracy with an increment of 7.54%. The successfully experimented spectral measures after training, optimizing, and testing were deployed for examining effective

FIGURE A2.26 Classified cotton by SIDSAM-tan at α-cut = 0.8.

identification of cotton crop from a homogeneous multispectral image (Figure A2.26). It is noted that the spectral information divergence measure efficiently identified cotton crops with overall accuracy (84.95%) and a kappa coefficient (0.7916).

CASE STUDY 9: KERNEL BASED PCM CLASSIFICATION APPROACH

Since the linear-PCM classifier was not able to handle the mixed pixel problem and non-linearity in the data adequately, thus, in order to handle the mixed pixel problem and non-linearity, the kernel functions are incorporated with PCM classifier in this case study. Nine different kernel functions were incorporated with PCM classifier, and the fuzzy parameter was optimized for them.

Objectives of the Case Study

The main objective of this case study is to develop a method to separate the classes having non-linear boundaries using KPCM. The specific objectives are:

1. To develop an objective function for kernel based PCM (KPCM) classifier.
2. To derive a method for selecting parameters for optimal kernel function.
3. To evaluate the performance of developed KPCM classifier in case of untrained classes.
4. To study the performance of single/composite kernels with PCM classifier.
5. To compare the performance of PCM with the developed KPCM classifier.

Study Area and Data Used

The geographical location of the study area is identical to Case Study 1. The acquisition and other details of satellite data are identical to Case Study 5. Simulated images of Formosat-2 as in Case Study 1 (Figure A2.2) have been used to study the behavior of developed KPCM algorithm accurately.

Methodology

The supervised KPCM classifier (Chapter 5) was developed with an aim to handle non-linearity between the classes. In this step, the best kernel function selected from nine different kernel functions was incorporated into PCM. The optimized value for fuzzy parameter was used for classification. The steps followed in supervised classification using KPCM classifier were as follows (Richards, 1993) (Figure A2.27):

1. Identifying the required land cover classes into which the image has to be classified. This ground cover data was used for training the classifier and evaluating the accuracy.
2. Identifying the ground data in the image for each class. This data is known as training data. Training data was collected through ground surveys and the photointerpretation method.
3. Using the training data to estimate the parameter for KPCM. These parameters are known as signature of the class. This step is known as training of classifier.
4. Using the trained KPCM classifier to calculate the membership value of feature vectors for each class. These per class classified maps are known as thematic maps.
5. Using the higher resolution classified results as referenced data (Formosat-2) for computing accuracy of the classification.

Results

The hyper tangent kernel (Chapter 5) was identified as the best performing kernel function as it showed highest overall accuracy of 98.37% and low entropy value of 0.48 as compared to linear PCM classifier, which showed low overall accuracy of 78.38% and high entropy of 0.5430. The better classification with KPCM classifier for mixed pixel was achieved with the classification of simulated images. To add the best outcome from different kernels, the composite kernel was formed by fusing the best performing hyper tangent kernel and sigmoid kernel using the weighted summation approach, and the value of weight constant was also optimized for composite kernel. The accuracy assessment results for the composite kernel were similar to the best performing hyper tangent kernel. An improved average user's accuracy of 89.90% was obtained with composite kernel, whereas the average user's accuracy with KPCM classifier was 89.17%. Hyper tangent KPCM classification was unaffected in the presence of untrained classes as compared to PCM classification by showing a very negligible effect in correlation values. The results revealed that the hyper tangent KPCM was consistently performing better with Landsat-8 data as well as with Formosat-2 data in the presence of non-linearity as well as in the absence of non-linearity.

As shown in Figure A2.28, the highest overall accuracy of 98.87% was achieved with hyper tangent kernel function for fuzzy parameter equal to 2.7. The second best result was observed with the sigmoid kernel with maximum overall accuracy of 92.60% at fuzzy parameter value equal to 1.5.

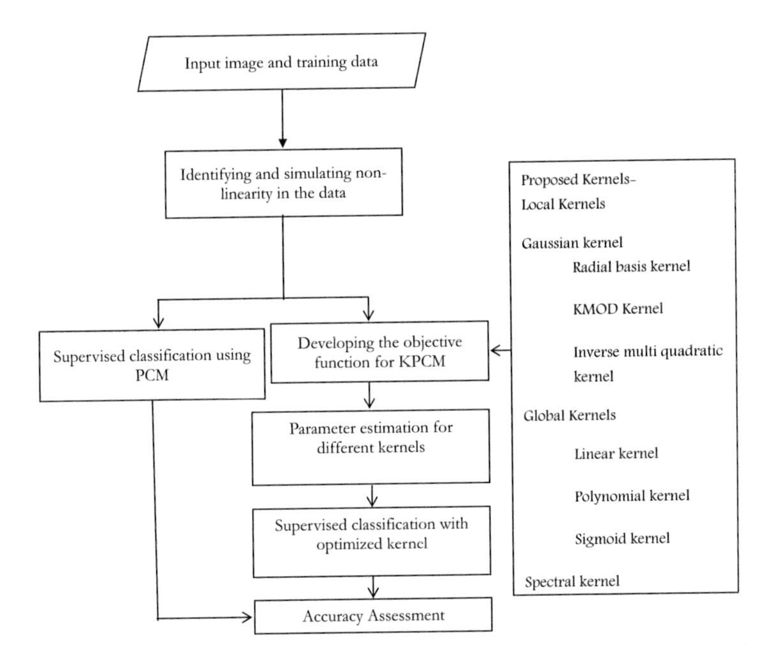

FIGURE A2.27 Methodology adopted.

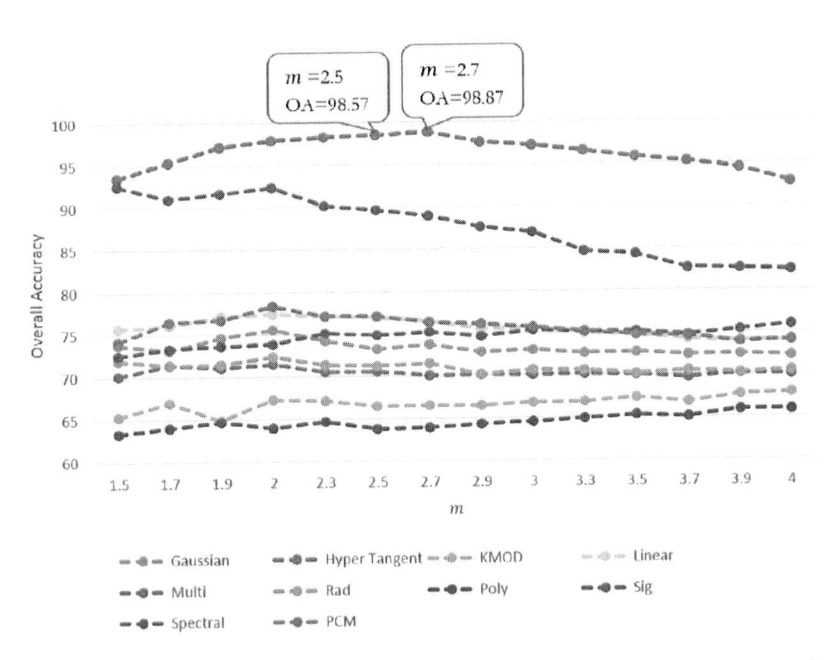

FIGURE A2.28 Overall accuracy of different kernels using FERM with respect to fuzzy parameter (m).

CASE STUDY 10: EFFECT OF RED EDGE BANDS IN FUZZY CLASSIFICATION: A CASE STUDY OF SUNFLOWER CROP

In this study the supervised modified possibilistic *c*-means (MPCM) classification approach has been adopted for the identification of sunflower fields due to the capability of handling outliers, noise, extraction of single crop, and coincident cluster problem. The classification approach was applied on four different modified temporal vegetation indices. The best vegetation index and suitable red-edge band for the discrimination of the sunflower crop were determined. Further, optimization of temporal date images to separate mapping of early sown, middle sown, and late sown fields was also identified. From the results of this study it has been proven that for temporal datasets, red-edge based indices are better than the standard indices for distinguishing between different crops while applying the MPCM classification method.

The supervised modified possibilistic *c*-means (MPCM) classification approach was adopted for the identification of sunflower fields that can deal with outliers, noises, single crop extraction, and coincident cluster problems. The classification was done with four different modified vegetation indices. The modified vegetation indices are generated by taking different combinations of red and red edge reflectance in a controlled manner. The value of the fuzzy weight constant and the weight of red edge bands in the calculation of modified vegetation indices are optimized. The best vegetation index and suitable red edge band with corresponding weight for the discrimination of sunflower crops are determined. The output assessment is done through the mean membership difference (MMD) method taken in two ways: one taken between early sown sunflower crop and wheat and the other between early sown, middle sown, and late sown sunflower crop. Optimization of temporal images for the separate mapping of early sown, middle sown, and late sown fields is also conducted. From the results obtained it is proven that for temporal datasets, the modified indices generated with red edge bands are better than the standard indices for distinguishing between different crops using the MPCM classification method.

Objectives of the Case Study

Objectives of this study are defined as follows:

1. To identify the best vegetation index for temporal database generation for MPCM classification.
2. To identify the suitable red edge band and corresponding weight constant for crop mapping.
3. To optimize the temporal images for early sown, middle sown, and late sown field identification.

Study Area and Data Used

The study area (Figure A2.29) considered for this research work is Shahabad, one of the three tehsils in Kurukshetra district, Haryana, India. Haryana state is in the northern part of India and contributes a major share in the agriculture production

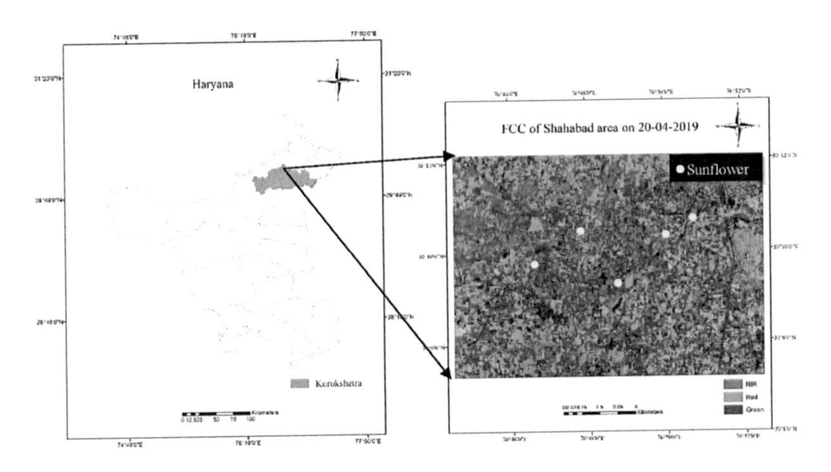

FIGURE A2.29 Study area: Shahabad, Haryana, India.

of wheat and rice. Shahabad (30° 10′ 6″ N, 76° 52′ 12″ E) lies between the two major cities Kurukshetra and Ambala. It is situated on a fertile riverbank which is mainly used for wheat, sunflower, and vegetable cultivation. A field visit of the study area was conducted on April 10, 2019, for identification of reference locations of sunflower crops (Figures A2.30–A2.32). The major agriculture varieties during the Rabi are season are sunflower and wheat. Sunflower fields in different growing stages were located and grouped into early sown (S1), middle sown (S2), and late sown (S3) categories.

FIGURE A2.30 Early sown sunflower field in Nalvi village.

FIGURE A2.31 Middle sown sunflower field near Shahabad Markanda.

FIGURE A2.32 Late sown and early sown sunflower field near Jharouli village.

TABLE A2.8
Temporal Datasets Used for the Study

Sentinel-2A (L2A product)	Sentinel-2B (L2A product)
March 21, 2019	March 3, 2019
March 31, 2019	April 5, 2019
April 2, 2019	April 15, 2019
April 30, 2019	May 5, 2019

The combined use of Sentinel-2A and -2B increases the temporal availability of datasets for the study area. Therefore, the temporal dataset for the study was acquired from multi-spectral imager (MSI) of Sentinel-2A and -2B satellites (L2A product). The MSI sensor contains 13 bands in which three are red edge bands with 20 m resolution. The dataset from March 21, 2019, is used in the study based on sowing period information of early sown sunflower crop collected during the field visit (Table A2.8).

Wheat fields were mainly at their harvesting period with some exceptions. Images of the sunflower fields taken during the field visit are shown in Figures A2.30–A2.32.

Methodology

The methodology adopted for the study is described in Figure A2.33. The temporal data (Table A2.8) acquired for the study contains different bands with different spatial resolution. In pre-processing of the data, all the bands required for the analysis are resampled to the identical pixel spacing of 10 m. The resampled dataset is given as input for the generation of the temporal indices database. Vegetation

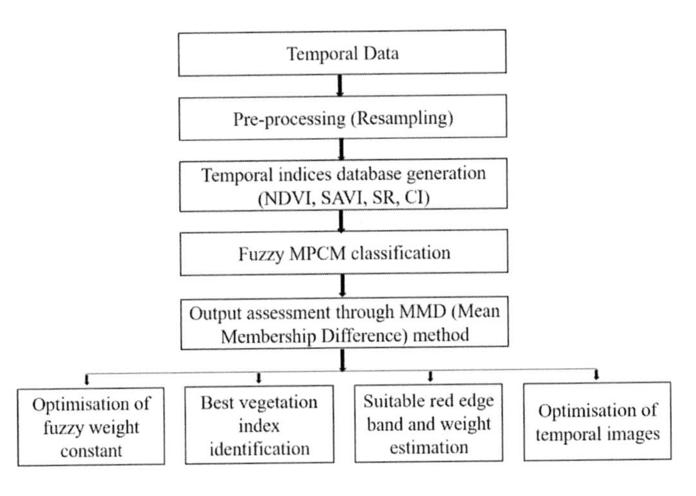

FIGURE A2.33 Methodology adopted for the study.

indices NDVI, SAVI, SR, and CI were generated with different compositions of red and red edge bands and are stacked together for the dates concerned for each particular band combination. The indices values have been calculated by varying the controlling parameter "a" from 0 to 1. The "a" value corresponding to minimum mean membership difference (MMD) was considered as an optimized controlling parameter value. The index corresponding to the optimized value of "a" from different temporal dates was then considered for making temporal indices, an input for the classifier.

The training sites were collected manually according to the 10n rule. The signature for classification was given from the generated training data accordingly for each classification input. The classification was done by varying the fuzzy weight constant from 1.2 to 3.0 with modified NDVI as input. The optimized value of '*m*' for other indices has been taken similar as for NDVI with an assumption that the dependency will be similar in all cases. MMD has been used for the assessment of outputs in terms of identifying the best vegetation index and control parameter, weighting exponent, suitable red-edge band, and optimization of temporal images.

Results

The effect of red edge bands in single crop identification using the fuzzy classification concept was analyzed by taking a sunflower crop as a case study. On comparison of the classified outputs based on sunflower and wheat discrimination, it was summarized that the suitable red edge band required for the accurate extraction of sunflower crop was different for different vegetation indices used. From the results obtained, it was evident that for temporal datasets the modified indices generated with red edge bands are better than the standard indices for distinguishing different crops using the MPCM classification method, while for discriminating between different growth stages of the same crop, the effect of red edge bands were found to be less productive. The outputs are shown in Figures A2.34–A2.36.

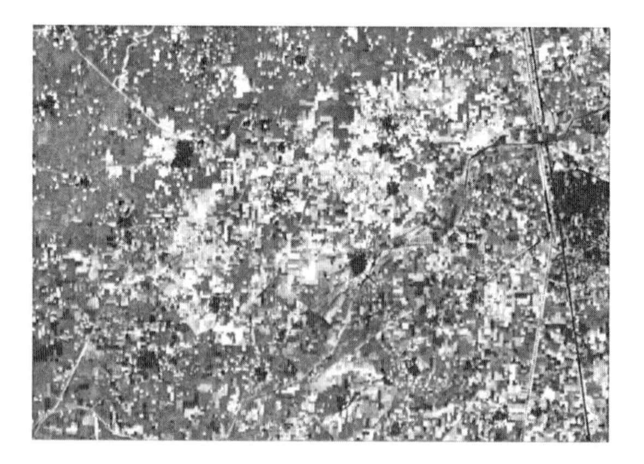

FIGURE A2.34 The classified output obtained for modified NDVI with red edge band 1 ("a" = 0.7).

FIGURE A2.35 The classified output obtained for modified SAVI with red edge band 1 ("a" = 0.7).

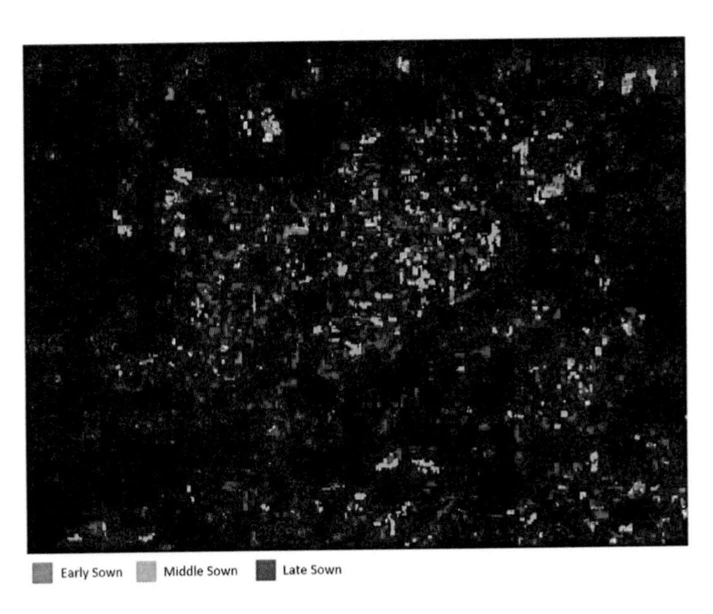

FIGURE A2.36 Stacked classified output obtained for modified SAVI where red represents S1, green represents S2 and blue represents S3 sunflower fields.

CASE STUDY 11: DISCRIMINATING SUGAR RATOON/PLANT CROP USING MULTI-SENSOR TEMPORAL DATA

The objective of this research work was to discriminate sugarcane ratoon/plants that are of interest to sugar mill industries or government agencies for better decision making processes. Sugarcane in India is a high priority crop for the government given the fact that India is the second largest producer of sugar in the world and the largest consumer of the sugar produced in the world. Any shortage in the sugar produce would have large impact on the sugarcane industry and also on the economy of the country in the end. Hence there is a need to prepare specific crop maps in order to be well equipped for any shortage in agricultural produce. The need of temporal data for continuous monitoring of crops and the unavailability of continuous temporal data is a well-known problem. So, data from different optical sensors like LISS-III and AWiFS (from IRS-P6) and TM from Landsat-5 was used to try to solve this problem.

Objectives of the Case Study

The main objective of this study is to discriminate a specific crop using a temporal single and multi-sensor data approach. This objective can be met by the following actions:

1. To identify crop spectral growth profile using a temporal and multi-sensor approach.
2. To study the separability between the target crop and other crops or vegetation based on their spectral growth profile.
3. To classify mixed pixels using fuzzy PCM technique.
4. To investigate the accuracy of the fuzzy classification method adopted with few operators like MIN, LEAST, and PROD.

Study Area and Data Used

The study area selected for this research work was Deoband city in Saharanpur district of the state of Uttar Pradesh, India. It is located in the upper doab region of Uttar Pradesh. This city lies to the northern part of India with a center latitude and longitude of 29.620 N and 77.670 E, respectively. The study area spans about 15 km in the east–west direction and about 6 km in the north–south direction. The main crops grown in this area are sugarcane, wheat, and plantation crops like mango, poplar, etc. The many sugarcane processing mills present in this area procure the bulk of the sugarcane produce providing a steady source of income for the farmers and also help in generating employment for the many seasonal agricultural laborers. The abundance of sugarcane farming and ready availability of data were the main reasons for the selection of the area for this study. This study deals with the multi-sensor data like LISS III and AWiFS from Resourcesat-1 satellite and TM sensor aboard Landsat-5 satellite.

Methodology

The flowchart of the methodology followed for this case study is shown in Figure A2.37. The detailed explanation of the methodology followed for discrimination of sugarcane ratoon/plant crops using the temporal single and multi-sensor approach is given in the following sections.

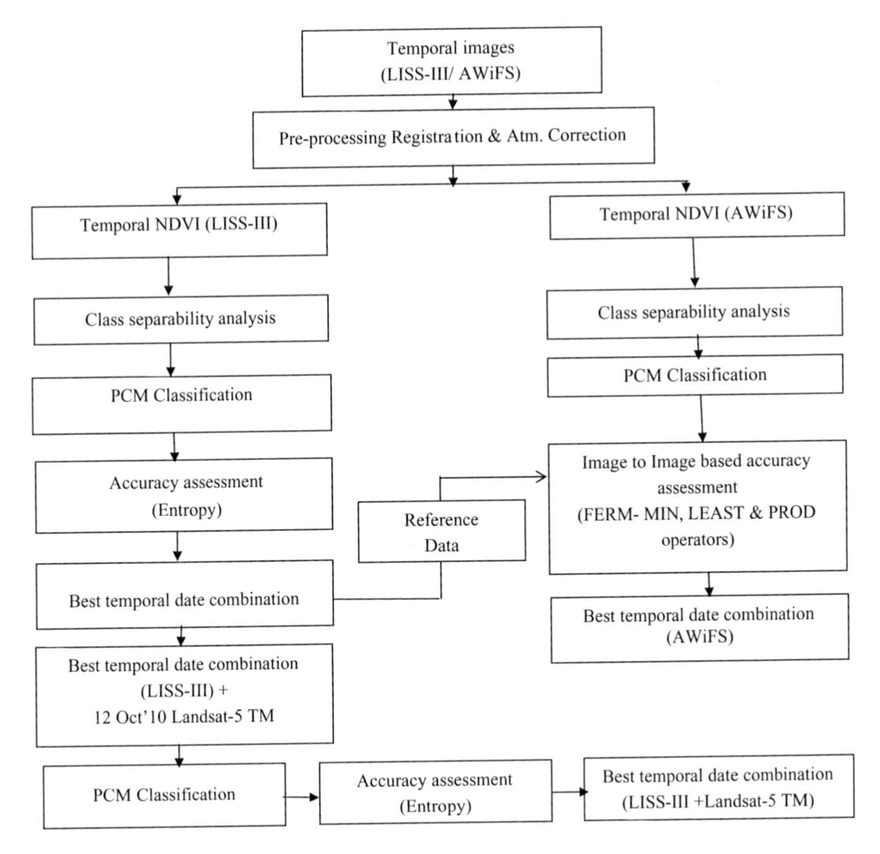

FIGURE A2.37 Flowchart showing methodology for the temporal and multi-sensor sugarcane ratoon/plant crop discrimination study.

Results

For an accurate estimation of area, PCM (possibilistic *c*-means), a possibilistic fuzzy based classifier capable of extracting single class in an image, was used. A spectral separability analysis (using single sensor data from LISS-III and AWiFS separately) was conducted between the class of interest (sugarcane plant and ratoon) and the non-interest classes to select the best 2, 3, 4 … date combinations to discriminate the class of interest. Combinations of these best dates were then classified using PCM classifier to extract sugarcane ratoon/plant to find the best overall date combination to discriminate sugarcane ratoon/plant. In the absence of any reference data, the soft classified outputs from the LISS-III sensor were assessed using an entropy measure criterion. The date combinations providing the least entropy was selected as the optimum date combination for discriminating the specific class. This date combination from LISS-III was used as a reference for assessing the soft classified outputs from AWiFS sensor using an image-to-image accuracy assessment technique. Various operators like MIN, LEAST, and PROD were also evaluated for their behavior and effectiveness in image-to-image accuracy assessment. In the second case, the effect of data from another sensor, i.e.,

FIGURE A2.38 Plant and ratoon sugarcane mapping from temporal LISS-III data.

Landsat-5 TM, when added to the optimum date combination from LISS-III was also evaluated. It was found that the entropy of the classified outputs from the selected best dates combination and multi-sensor approach was lower than the entropy measured from the single sensor (LISS-III) approach. Lower entropy means lesser uncertainty associated with classification and accuracy was higher, and vice-versa. This study explored the applicability of temporal single and multi-sensor data for discrimination of specific crops, sugarcane plant and ratoon (Figure A2.38). A multi-sensor approach helped in increasing the temporal data sampling for the continuous monitoring of crops when data available from any single sensor approach was insufficient. The end result of this study was the benefit of using the right temporal dates for discriminating sugarcane plant and ratoon. Such information is developed skillfully by agricultural scientists in selecting an optimum number of strategically placed temporal images in the crop growing season for discriminating the specific crop accurately.

Machine Learning or Deep Learning or Human Learning Happens Through Training.....

....Comes only from....

Practice....

Practice....

Practice....

Index

Note: Page numbers in bold and italics refer to tables and figures, respectively.

N

O

P

T

U

V

W

for taken 'me' in your hands